Novel Imaging and Spectroscopy

Edited by Jinfeng Yang

Published in London, United Kingdom

IntechOpen

Supporting open minds since 2005

Novel Imaging and Spectroscopy
http://dx.doi.org/10.5772/intechopen.83198
Edited by Jinfeng Yang

Contributors
Hiroaki Matsui, Luigi Sirleto, Rajeev Rnajan, Maria Antonietta Ferrara, Jinfeng Yang, Hirohide Serizawa, Boucerredj Noureddine, Beggas Khaled, Haider Al-Tameemi, Bence Tamás Szabó, Adrienn Dobai, Csaba Dobo-Nagy, Koichi Kan, Masao Gohdo, Yoichi Yoshida, Hidehiro Yasuda

First published in London, United Kingdom, 2020 by IntechOpen
IntechOpen is the global imprint of INTECHOPEN LIMITED, registered in England and Wales, registration number: 11086078, 7th floor, 10 Lower Thames Street, London, EC3R 6AF, United Kingdom
Printed in Croatia

British Library Cataloguing-in-Publication Data
A catalogue record for this book is available from the British Library

Additional hard and PDF copies can be obtained from orders@intechopen.com

Novel Imaging and Spectroscopy
Edited by Jinfeng Yang
p. cm.
Print ISBN 978-1-83880-051-2
Online ISBN 978-1-83880-052-9
eBook (PDF) ISBN 978-1-83880-914-0

For EU product safety concerns:
IN TECH d.o.o., Prolaz Marije Krucifikse Kozulić 3, 51000 Rijeka, Croatia, info@intechopen.com or visit our website at intechopen.com.

Meet the editor

Jinfeng Yang is an Associate Professor of accelerator physics and materials science at the Institute of Scientific and Industrial Research, Osaka University, Japan. He has worked on the generation and applications of high-brightness femtosecond/picosecond pulsed electron beams in the particle accelerator field for over twenty years. He has published more than 100 papers in indexed scientific journals and international conferences. He developed a pulse radiolysis with femtosecond resolution. The development opened the first experimental study of radiation chemistry in femtosecond time regions. His current research is concerned with the development of ultrafast electron diffraction/microscopy with relativistic femtosecond electron pulses and the study of ultrafast phenomena including structural dynamics and chemical/biochemical reactions.

Contents

Preface

Since the birth of light microscopy, various imaging and spectroscopic techniques, including electron microscopy, X-ray imaging, and absorption/emission spectroscopy, have been developed. The technologies have played an important role in physics, chemistry, and life science. Electron microscopy and X-ray imaging have been applied to directly observe three-dimensional (3D) material structures at atomic scales. Spectroscopies have been used to detect, identify, and quantify information on atoms and molecules. Research using imaging and spectroscopic techniques have brought abundant information on material culture to mankind. Many significant physics and chemical laws have been constructed through measurements of how materials respond in these experiments, but to truly understand what is going on, more sophisticated apparatus would be needed.

The material properties or dynamic phenomena we observe on macroscopic scales result from the countless interactions that take place between individual atoms on timescales as fast as a picosecond or femtosecond. For example, the OH stretch of water has a period of 10 femtoseconds. The motions involved are less than 0.1 angstrom. To study the processes on such intricate scales, time-resolved spectroscopies using femtosecond-pulsed lasers were proposed in the 20th century. At the beginning of the 21st century, ultrafast imaging spectroscopy with ultrashort-pulsed electrons and X-rays were adapted using real-time and real-space imaging of dynamical processes in matter. Many transient phenomena were revealed, including dynamics of photodissociation and chemical reactions, photon-induced lattice heating and melting on picosecond time scales, other structural phase transitions, etc. The recent developments have opened the femtosecond time domain to atomically resolved dynamics.

Medical imaging is an indispensable imaging technology in our life. It is undergoing a revolution from analog imaging to digital imaging and has shifted from general X-ray radiographs to new modalities such as computerized tomography (CT), magnetic resonance imaging (MRI), and isotope imaging. CT, which combines the power of computer processing with X-ray imaging, provides high-resolution images of the bony structures in three different planes. MRI acquires images of internal body structures and becomes the imaging modality of choice for soft tissues and vascular structures. Isotope imaging is applied in the elucidation of hidden causes of pain such as tumors or cancers.

In this book, we introduce several novel imaging and spectroscope techniques and their applications concerning such subjects:

- In Chapter 1, a 4D imaging technique with relativistic femtosecond electron pulses is reviewed. It is also ultra-high voltage pulsed electron microscopy and is used for femtosecond atomic imaging with single shot.

- In Chapter 2, a methodology of single-shot time-resolved diffraction imaging with an excellent temporal resolution of femtoseconds is reported for the study of ultrafast dynamics of photo-induced irreversible phase transitions.

- In Chapter 3, a femtosecond pulse radiolysis (femtosecond time-resolved spectroscopy) is reviewed. The observation of femtosecond/picosecond kinetics and reactions of hydrated and pre-hydrated electrons in water is reported.

- In Chapter 4, the observation of surface plasmons and optical dynamics of vanadium dioxide on femtosecond and picosecond time scales using vis-NIR and FTIR spectrometers based on a femtosecond laser system is reported.

- In Chapter 5, the design and implementation of 2D stimulated Raman microscopy using femtosecond lasers are described. The imaging of polystyrene beads and biological samples are shown.

- In Chapter 6, an interesting diffraction phenomenon of an electromagnetic plane wave by a rectangular hole is introduced.

- In Chapter 7, a study of nanoplasma formation of Na and Kr clusters irradiated by intense femtosecond laser is reported. A modified nanoplasm model is given to examine the cluster explosion dynamics.

- Chapter 8 reviews the advantages and limitations of the most important medical imaging techniques, including plan X-ray radiograph, magnetic resonance imaging, computerized tomography, isotope imaging, etc.

- In Chapter 9, a cone-beam computed tomography in dentomaxillofacial radiology is reviewed. Its current limitations and expected improvements are described.

Although this book includes a limited number of topics, I think that the content in each chapter will be impressive to the reader. I hope this book will contribute to future developments and applications.

Finally, I am grateful to all authors for their contributions to this book and their efforts to complete the chapters. I also acknowledge the IntechOpen publishing team, especially Mateo Pulko for cooperation in the publishing process.

Dr. Jinfeng Yang
The Institute of Scientific and Industrial Research,
Osaka University,
Japan

Chapter 1

Introductory Chapter: 4D Imaging

Jinfeng Yang and Hidehiro Yasuda

1. Introduction

The study of ultrafast phenomena, including structural dynamics and molecular reactions, is of great interest for physics, chemistry, biology, and materials science. There are numerous examples of phase transitions in condensed materials and chemical reactions in free molecules proceeding on nanosecond, picosecond, and even femtosecond time scales. To study processes or reactions on such intricate scales, more sophisticated apparatus would be needed. It is well known that electron microscopy is a powerful imaging technique and is applied to a wide research field. The progress of electron microscopy has shown that three-dimensional (3D) material structures can be observed with an atomic spatial resolution. However, the conventional electron microscopy does not allow studying ultrafast processes because of the limitation of the speed of video camera.

The study of ultrafast structural dynamics or molecular reactions requires the use of probes ensuring not only high spatial but also high temporal resolutions. For this purpose, the new development of ultrafast electron microscopy (UEM), by combining temporal resolution into conventional electron microscopy, has been begun in the world. UEM uses a short pulsed electron beam replacing the continuous electron beam in the conventional electron microscopy to image the atomic motion by time-resolved recording in real time. By introducing temporal resolution into 3D electron microscopy, UEM allows us to observe the four fundamental dimension structures of matter: three spatial and one temporal, which is called 4D imaging.

Recent developments in UEM have shown that spatial and temporal information of matter can be obtained simultaneously on very small and fast scales. The first UEM was proposed to observe fast processes using a modified 120-keV electron microscope by Ahmed H. Zewail, Nobel Prize winner in Chemistry 1999, in the California Institute of Technology [1, 2]. He and his colleagues succeeded to observe the laser-photon-induced picosecond structural phase transition in vanadium dioxide film using a stroboscopic method with "single" electron pulses [3]. Later, a hybrid 200-keV apparatus was developed. A spatial-temporal resolution of 3.4 Å and 250 fs has been achieved. Recently, there are many research activities focused on improving the electron source and electron optics inside the microscope to achieve better temporal and spatial resolutions [4–9]. However, in the current UEM, the samples must be pumped 10^7 times or more by the laser. The process being studied must be perfectly reversible. To study the irreversible processes, it is necessary to record images with a larger number of electrons per pulse possible.

In this chapter, we introduce a novel UEM method with relativistic-energy electron pulses. In this relativistic UEM, an advanced radio-frequency (rf) acceleration technology is used to generate relativistic femtosecond electron pulses containing a large number of electrons in pulse and to achieve single-shot femtosecond imaging for the study of ultrafast irreversible structural processes.

2. UEM with relativistic femtosecond electron pulses

The relativistic UEM [10–14] is constructed with three principal components: a rf acceleration-based electron gun, a condenser system, and an imaging system. **Figure 1** shows a photo of the relativistic UEM, which is 3.5 m in height and 0.8 m in diameter. The rf electron gun is driven by a high power of rf to generate a high-peak rf electric field of 100 MV/m, which is 10 times higher than that of direct current gun in the conventional electron microscopy. The electrons emitted from photocathode are then quickly accelerated by the rf electric field into the relativistic energy region to reduce the effect of space charge, yielding ultrashort pulses containing a large number of electrons in pulse. The details of the rf electron gun and the generation of femsecond electron pulses are described in Chapter 2 [15].

Next, the electrons pass through a series of condenser lenses, which use magnetic field to precisely control the intensity of the beam, and its illumination angle on the sample. A relativistic-energy electron imaging system, including an objective lens, an intermediate lens and two projector lenses, is used to magnify the microscopic images. Finally, the images are recorded with a viewing screen (scintillator)

Figure 1.
Photo of UEM with relativistic-energy femtosecond electron pulses constructed at Osaka University [13, 14].

via a charge-coupled device camera [16]. The relativistic UEM is also an ultra-high voltage transimission electron microscopy (TEM). It exhibits many significant advantages over nonrelativistic-energy UEMs:

1. High temporal resolution of 100 fs or less is achievable, because the ultrashort electron pulses of <100 fs can be produced by the rf gun. The transit-time broadening due to the relative energy spread is reduced using the relativistic-energy electrons.

2. The relativistic UEM enables to observe the irreversible processes in materials by single-shot imaging with high-intensity femtosecond electron pulses.

3. The high-energy electrons significantly increase the extinction distance of elastic scattering. Our previous studies [17, 18] indicate that the kinematic theory with the assumption of single elastic scattering events can be applied in the relativistic UEM. This enables one to easily explain structural dynamics from the experimental results.

4. A thick sample can be used for measurement, thus obviating the requirement to prepare suitable thin samples.

5. The relativistic UEM is suitable for in situ observations. A large pole piece of the objective lens can be applied for installing various specimens.

The structural dynamics is observed in UEM with a pump-and-probe method, as shown in **Figure 2**. The femtosecond laser pulse is used as a pump pulse to excite the sample, while the electron pulse is used to record the time evolution of image of the structure by changing the time interval between the electron pulse and the laser pump pulse. The time resolution of UEM is determined mainly by the pulse duration s of the probe electrons and the pump laser. A high temporal resolution can be achieved with the ultrashort electron pulse and the ultrashort laser pump pulse. In this UEM, many demonstrations have been carried out and summarize as the followings:

1. A 100-fs-long pulsed beam containing 10^6–10^7 electrons at an energy of 3 MeV has been generated using the rf gun [10–12].

2. In the imaging experiments using these femtosecond pulses, we successfully observed contrast TEM images of 200-nm-diameter gold nanoparticles and other materials. At a low-magnification observation, single-shot imaging with the 3 MeV fs electron pulse is achievable [11, 12].

3. In the electron diffraction measurement, we successfully detected high-contrast electron diffraction images of single crystalline, polycrystalline, and amorphous materials. An excellent spatial resolution of diffraction images was obtained as 0.027 $Å^{-1}$ [19, 20].

4. In the pump-and-probe experiments using the relativistic femtosecond pulses, a laser-induced ultrafast melting dynamics in crystalline gold [17, 18] and a laser-excited ultrafast electronically driven phase transition in single-crystalline silicon [19, 20] were observed. The best temporal resolution of 100 fs has been achieved [20].

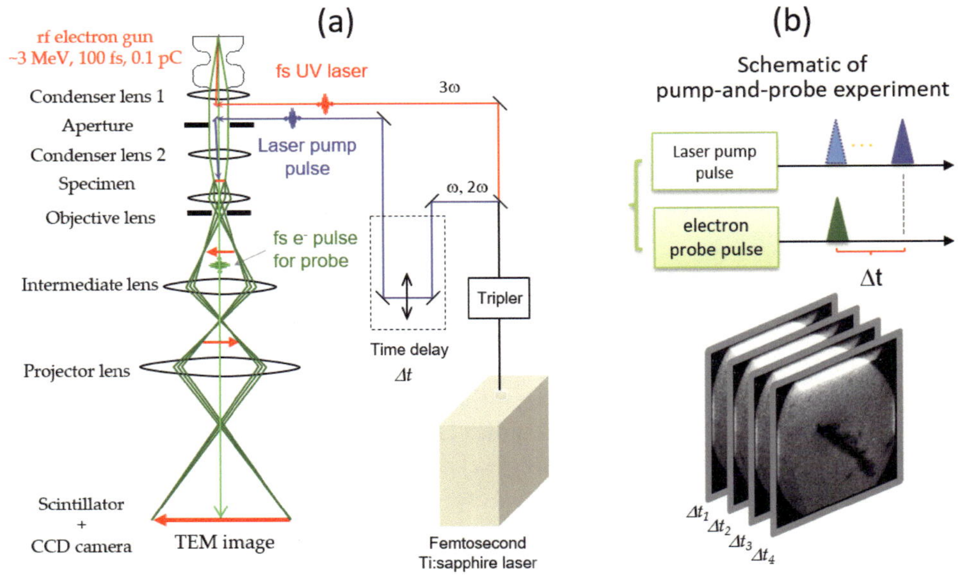

Figure 2.
(a) General schematic of UEM using relativistic femtosecond electron pulse and (b) pump-and-probe method for the observation of structural dynamics [13].

The details of the above experiments have been reported in the related references. The results demonstrate the advantages of relativistic UEM, including access to high-order Bragg reflections, single-shot imaging with the relativistic femtosecond electron pulse, and the feasibility of time-resolved imaging to study ultrafast structural dynamics.

3. Conclusion

Ultrafast electron microscopy with relativistic femtosecond electron pulses is a very promising 4D imaging technique for scientists wishing to study ultrafast structural dynamics in materials. It is an unprecedented innovative technology that enables femtosecond atomic-scale imaging using single-shot measurement and paves the way for the study of irreversible processes in physics, chemistry, biology, and materials science.

The relativistic UEM is also a very compact, ultra-high voltage electron microscopy. It can be used in a variety of settings such as general research institutions and laboratories. Furthermore, by providing a femtosecond temporal resolution, the relativistic UEM will constitute the next generation of electron microscopes. It will allow the study of structural dynamics to be broken into unprecedented timeframes, further encouraging the discovery of new knowledge.

Acknowledgements

The authors acknowledge Prof. Yoshida Y., Drs. Kan K. and Gohdo M. of the Institute of Scientific and Industrial Research in Osaka University for their valuable discussions, Profs Tanimura K. of the Research Center for Ultra-High Voltage Electron Microscopy (UHVEM) in Osaka University for their valuable suggestions. Additionally, the authors thank Urakawa J., Takatomi T., and

Terunuma N. of the High Energy Accelerator Research Organization (KEK) for the fabrication of the rf gun.

This research was funded by JSPS KAKENHI Grant Numbers JP22246127, JP26246026, and JP17H01060 of Grant-in-Aid for Scientific Research (A) and JP16K13687 of Challenging Research Exploratory, Japan.

Author details

Jinfeng Yang[1*] and Hidehiro Yasuda[2]

1 The Institute of Scientific and Industrial Research, Osaka University, Osaka, Japan

2 Research Center for Ultra-High Voltage Electron Microscopy, Osaka University, Japan

*Address all correspondence to: yang@sanken.osaka-u.ac.jp

References

[1] Zewail AH, Thomas JM. 4D Electron Microscopy: Imaging in Space and Time. London: Imperial College Press; 2010. DOI: 10.1142/p641

[2] Zewail AH. Four-Dimensional Electron Microscopy. Science. 2010;**328**:187-193. DOI: 10.1126/science.1166135

[3] Grinolds MS, Lobastov VA, Weissenrieder J, Zewail AH. Proceedings of the National Academy of Sciences of the United States of America. 2006;**103**:18427-18431. DOI: 10.1073/pnas.0609233103

[4] Piazza L, Masiel DJ, LaGrange T, Reed BW, Barwick B, Carbone F. Design and implementation of a fs-resolved transmission electron microscope based on thermionic gun technology. Chemical Physics. 2013;**423**:79-84. DOI: 10.1016/j.chemphys.2013.06.026

[5] Bücker K, Picher M, Crégut O, LaGrange T, Reed BW, Park ST, et al. Electron beam dynamics in an ultrafast transmission electron microscope with Wehnelt electrode. Ultramicroscopy. 2016;**171**:8-18. DOI: 10.1016/j.ultramic.2016.08.014

[6] Feist A, Bach N, Rubiano da Silva N, Danz T, Möller M, Priebe KE, et al. Ultrafast transmission electron microscopy using a laser-driven field emitter: Femtosecond resolution with a high coherence electron beam. Ultramicroscopy. 2017;**176**:63-73. DOI: 10.1016/j.ultramic.2016.12.005

[7] Kuwahara M, Nambo Y, Aoki K, Sameshima K, Jin X, Ujihara T, et al. The Boersch effect in a picosecond pulsed electron beam emitted from a semiconductor photocathode. Applied Physics Letters. 2016;**109**:013108. DOI: 10.1063/1.4955457

[8] Houdellier F, Caruso GM, Weber S, Kociak M, Arbouet A. Development of a high brightness ultrafast transmission electron microscope based on a laser-driven cold field emission source. Ultramicroscopy. 2018;**186**:128-138. DOI: 10.1016/j.ultramic.2017.12.015

[9] Manz S, Casandruc A, Zhang D, Zhong Y, Loch RA, Marx A, et al. Mapping atomic motions with ultrabright electrons: Towards fundamental limits in space-time resolution. Faraday Discussions. 2015;**177**:467-491. DOI: 10.1039/C4FD00204K

[10] Yang J, Yoshida Y, Shibata H. Femtosecond time-resolved electron microscopy. Electronics and Communications in Japan. 2015;**98**: 50-57. DOI: 10.1002/ecj.11763

[11] Yang J, Yoshida Y, Yasuda H. Ultrafast electron microscopy with relativistic femtosecond electron pulses. Microscopy. 2018;**67**:291-295. DOI: 10.1093/jmicro/dfy032

[12] Yang J. Ultrafast electron microscopy with relativistic femtosecond electron pulses. In: Arita M, Sakaguchi N, editors. Electron Microscopy: Novel Microscopy Trends. Croatia: IntechOpen; 2019. DOI: 10.5772/intechopen.81405

[13] Yang J. Ultrafast electron microscopy reinventing femto-second atomic scale imaging. Research OUTREACH. 2020;**112**:26-29. DOI: 10.32907/RO-112-2629. Available from: https://researchoutreach.org/articles/ultrafast-electron-microscopy-reinventingfemtosecond-atomic-scale-imaging/

[14] Yang J. New crystallography using relativistic femtosecond electron pulses. Impact. 2019;**10**:76-78. DOI: 10.21820/23987073.2019.10.76

[15] Yang J. Femtosecond electron diffraction with relativistic electron

pulses. In: Yang J, editor. Novel Imaging and Spectroscopy. Croatia: IntechOpen; 2020. DOI: 10.5772/intechopen.88511

[16] Murooka Y, Naruse N, Sakakihara S, Ishimaru M, Yang J, Tanimura K. Transmission-electron diffraction by MeV electron pulses. Applied Physics Letters. 2011;**98**:251903. DOI: 10.1063/1.3602314

[17] Giret Y, Naruse N, Daraszewicz SL, Murooka Y, Yang J, Duffy DM, et al. Determination of transient atomic structure of laser-excited materials from time-resolved diffraction data. Applied Physics Letters. 2013;**103**:253107. DOI: 10.1063/1.4847695

[18] Daraszewicz SL, Giret Y, Naruse N, Murooka Y, Yang J, Duffy DM, et al. Structural dynamics of laser-irradiated gold nanofilms. Physical Review B. 2013;**88**:184101. DOI: 10.1103/PhysRevB.88.184101

[19] Yang J, Yoshida Y. Relativistic ultrafast electron microscopy: Single-shot diffraction imaging with femtosecond electron pulses. Advances in Condensed Matter Physics. 2019;**2019**:9739241. DOI: 10.1155/2019/9739241

[20] Yang J, Gen K, Naruse N, Sakakihara S, Yoshida Y. A compact ultrafast electron diffractometer with relativistic femtosecond electron pulses. Quantum Beam Science. 2020;**4**:4. DOI: 10.3390/qubs4010004

Chapter 2

Femtosecond Electron Diffraction Using Relativistic Electron Pulses

Jinfeng Yang

Abstract

Observation of atomic-scale structural motion in matter with femtosecond temporal resolution is of considerable interest to scientists and paves the way for new science and applications. For this purpose, ultrafast electron diffraction (UED) imaging using femtosecond electron pulses is a very promising technique, as electrons have a larger elastic scattering cross section as compared to photons or X-rays and can be easily focused in observation with high spatial resolution. In this chapter, we first give an overview of the historical development of current nonrelativistic UEDs and discuss the potentials of UEDs with relativistic electron pulses. Second, we describe the concept and development of relativistic UED with femtosecond electron pulses generated by a radio-frequency acceleration-based photoemission gun. Some demonstrations of diffraction imaging of crystalline materials using 3-MeV electron pulses with durations of ~100 fs are presented. Finally, we report a methodology of single-shot time-resolved diffraction imaging for the study of ultrafast dynamics of photo-induced irreversible phase transitions.

Keywords: ultrafast electron diffraction, femtosecond electron pulse, relativistic electron beam, structural dynamics, radio-frequency electron gun

1. Introduction

Femtosecond imaging is a long-awaited technique for materials scientists to observe atomic and molecular motions directly and in real time. For this type of ultrafast imaging, time-resolved diffraction with short-pulsed X-rays has been developed and is widely used globally. Recently, ultrafast electron diffraction (UED) using femtosecond electron pulses [1–3] has facilitated the study of the structural dynamics of reversible and irreversible processes, including ultrafast phase transformations, femtosecond chemical/biochemical reactions, and radiation damages. It is well-known that electrons have many advantages over X-rays. Electrons have a larger elastic scattering cross section and can be easily focused to develop a measurement with high spatial resolution. A small sample can be used in the UED measurement. The instruments used with UEDs are also very compact.

The earliest time-resolved electron diffraction experiments with pulsed electrons on the milli- to microsecond time scale were developed using a deflection technique in the 1980s [4]. Later, a photoemission pulsed electron source was used in an electron diffraction measurement by Ewbank et al. [5, 6]. The temporal resolution was improved to sub-nanoseconds. In 1982, Mourou and Williamson [7] pioneered the use of 100-ps electron pulses to construct a picosecond electron diffraction experiment. Later, they used a streak-camera tube to generate 25-keV picosecond electron pulses (~20 ps) in a UED experiment and succeeded at detecting laser-induced ultrafast phase transformation in aluminum [8]. The number of electrons per pulse used in the diffraction measurement was maintained at $\sim 10^4$. Since 2000, the research groups of Zewail [1, 9, 10], Miller [2], and Cao [11] have focused their efforts on generating ultrashort electron pulses. A mode-locked femtosecond laser was used to generate ultrashort electron pulses and to excite the samples. A temporal resolution of sub-picoseconds and femtoseconds was achieved in UEDs, which provided real-time diffraction imaging and enabled the recording of atomic or molecular motion in chemical and biochemical reactions. **Figure 1** shows a schematic of a typical femtosecond electron-diffraction apparatus constructed with a laser-driven electron source, magnetic lens, sample-positioning system, and electron diffraction detector. In most UEDs, a conventional Ti:sapphire femtosecond pulsed laser is used to generate ultrashort electron pulses with a photocathode and to pump the sample to induce structural changes in materials. The key elements of this apparatus are the electron source and detection system.

The temporal resolution is mainly defined by the electron pulse duration. In all of the aforementioned UEDs, a DC acceleration-based photoemission electron gun was used to generate ultrashort electron pulses with a short-pulsed laser. The use of a higher accelerating electric field (extraction field, E_{ext}) between the photocathode and anode (grid) is crucial to reducing Coulomb repulsion of electrons (space-charge effects). The transit-time broadening τ_{KE} can be reduced by increasing E_{ext}, because $\tau_{KE} \propto \sqrt{\Delta E_{kin}}/E_{ext}$ [12], where ΔE_{kin} is the kinetic energy spread of photo-electrons. For an Au or Ag photocathode driven by a 266-nm ultraviolet (UV) laser, $\Delta E_{kin} \sim 0.6$ eV, the dispersive broadening near the photocathode can be maintained below 300 fs at $E_{\mathrm{ext}} = 10$ MV/m or more [3]. However, the maximum static extraction field in the dc gun is determined by the vacuum breakdown limit of ~10 MV/m.

The space-charge effects occur not only in the electron gun but also during the propagation of the electron pulse from the gun to the sample. The broadening of both the pulse duration and energy spread of electrons due to the space-charge effects in the propagation of nonrelativistic femtosecond electron pulses has been investigated theoretically by Siwick et al. [13]. Their results indicated that when

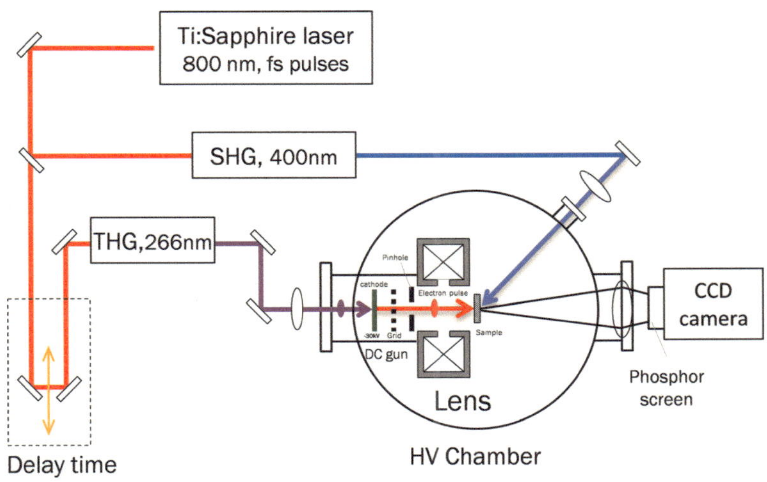

Figure 1.
Schematic of a typical femtosecond electron-diffraction apparatus. A DC-acceleration-based photoemission electron gun is used to produce ultrashort electron pulses with third harmonics of a Ti:sapphire femtosecond laser. The electron pulse passes through the specimen and produces a diffraction pattern of the structure.

a 30-keV and 300-fs electron pulse containing $\sim 10^4$ electrons is transported to the sample with a 40-cm-long drift space, the pulse duration is increased to 4 ps, and the relative energy spread is increased to 3×10^{-3}. In this case, both the temporal and spatial resolutions are reduced by the space-charge effects. To reduce the space-charge effects, two solutions are used in UED. One involves a decrease in the number of electrons in the pulse. This results in a low brightness electron beam at the sample and presents tremendous difficulties for single-shot imaging in the study of the dynamics of irreversible processes. Another involves reducing the photocathode-to-sample distance. Miller et al. at the University of Toronto developed a 30-keV electron gun with a photocathode-to-sample distance of 4.5 cm. This source produced a 600-fs short electron pulse containing 6000 electrons per pulse. The beam fluence at the sample $\varphi = N/A$, where N is the total number of electrons per pulse and A is the beam area at the sample, is $\varphi \sim 1 \times 10^{11}$ m^{-2} [14]. Nevertheless, a minimum of $\sim 10^4$–10^5 electrons per pulse are needed to resolve the diffraction peak and/or obtain a clear image in UED [15].

To generate high-brightness electron beams with pulse durations on the order of 100 fs, an advanced accelerator technology for radio-frequency (RF) acceleration-based photoemission electron guns (photocathode RF guns) has been proposed to generate multi-MeV femtosecond electron pulses for UED [16–27]. The RF gun is usually operated with a high RF electric field equal to or >100 MV/m. Therefore, the electrons emitted from the photocathode can be quickly accelerated into the relativistic energy region to minimize the space-charge effects in the pulse, yielding a femtosecond or picosecond pulse with numerous electrons. Recently, Yang et al. [28–30] developed the first prototype for relativistic ultrafast electron microscopy using the RF gun. They succeeded in generating high-brightness electron pulses with a pulse duration of 100 fs containing 10^7 electrons at an energy of 3 MeV. They also demonstrated the single-shot imaging using these femtosecond electron pulses [31, 32]. Relativistic UED is very promising for the study of ultrafast dynamics in solid-state materials and chemical/biological complex systems. It exhibits many crucial advantages over nonrelativistic UED systems:

- High-current electron pulses enable single-shot imaging so that both reversible and irreversible processes in materials can be observed.

- Higher energies considerably enhance the extinction distance for elastic scattering and provide structural information that is essentially free from multiple scattering and inelastic effects. This enables us to easily understand and explain structural dynamics.

- Ultrashort pulses of <100 fs are possible. The utilization of the relativistic electron pulse overcomes the loss of temporal resolution because of the velocity mismatch in samples. High temporal resolution of 100 fs or less can be achieved in relativistic UED.

- A thick sample can be used for measurement, thus obviating the requirement to prepare suitable thin samples.

- Radiation effects derived from ionization damage processes decrease at higher energies in samples. In fact, at MeV energies, the dominant damage process for electron beams is not ionization but rather the much slower ballistic “knock-on” process.

- Relativistic UED is suitable for in situ observations, as large areas exist in the sample room for installing various specimens. Relativistic UED can be used to study gas-, liquid-, and solid-phase samples.

In this chapter, we introduce UED with relativistic femtosecond electron pulses. The chapter also describes the generation of femtosecond electron pulses using an RF photoemission gun, the concept and design of relativistic UED, and demonstration experiments with relativistic femtosecond electron pulses.

2. Relativistic UED with femtosecond electron pulses

Relativistic UED consists of a 1.6-cell S-band (2.856 GHz) photocathode RF gun, an imaging system with magnetic lenses, a femtosecond laser, and a detector. **Figure 2** shows a schematic and photograph of relativistic UED at Osaka University. All components were installed on a vibration-controlled board with a size of $3 \times 3\ m^2$.

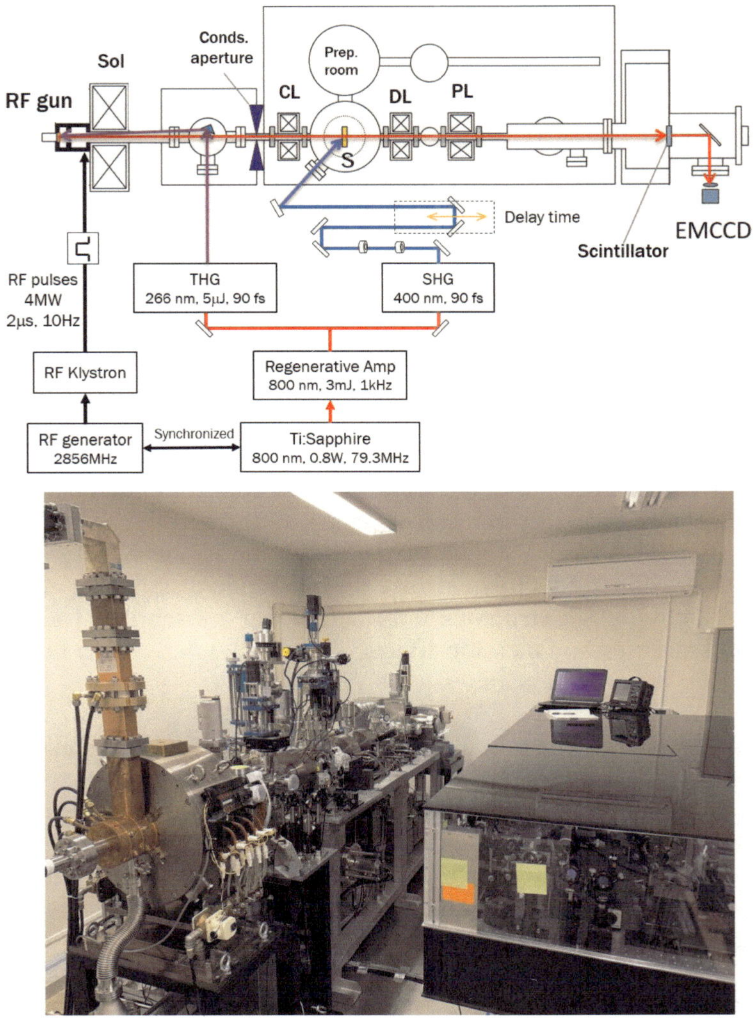

Figure 2.
Schematic (top) and photograph (bottom) of UED apparatus with relativistic femtosecond electron pulses at Osaka University. All components are installed on a vibration-controlled board with a size of $3 \times 3\ m^2$.

2.1 Photocathode RF gun

The photocathode RF gun is a high-brightness electron source and has been widely applied in the field of advanced particle accelerators (e.g., free-electron lasers and linear colliders). The RF gun used in relativistic UED consists of two RF cavities: a half cell and a full cell (1.6-cell), as shown in **Figure 3**. The length of the full-cell cavity is equal to half the 2.856 GHz RF wavelength, $\lambda/2$ = 52.48 mm, whereas the length of the half cell is 0.6 times $\lambda/2$; numerical studies have shown that an optimal performance is obtained if the half-cell cavity is 0.6 times the full cell length rather than 0.5×. The design and details of the cavities were described in [32, 33]. The cavities are driven by a MW-power 2.856 GHz accelerating RF to produce a stronger accelerating RF field on the photocathode. The RF gun is operated in the TM_{010} transverse magnetic mode [34, 35]. The phase shift between the half and full cells is equal to π, resulting in the acceleration of electrons in both cavities. The linear components of the electric fields at $r = 0$ (the center axis of the RF cavities) [32] can be assumed to be:

$$E_z = E_0 \cos kz \sin(\omega t + \varphi_0), \tag{1}$$

where E_0 is the peak accelerating field, $k = 2\pi/\lambda$, $\omega = ck$, c is the velocity of light, and ϕ_0 is the initial RF phase when the electron leaves the cathode surface ($z = 0$) at $t = 0$. In relativistic UED, the peak RF power filled to the RF cavities is 4 MW, resulting in E_0 = 75 MV/m. This field thus accelerates electrons emitted from the photocathode quickly up to 3 MeV to minimize the space-charge effects in the pulses. Therefore, the RF gun can easily produce femtosecond electron pulses by the illumination of the femtosecond laser on the photocathode.

The visualization of atomic-scale structural motion by UED requires electron pulses of the shortest duration and lowest emittance to achieve high temporal and spatial coherence. The temporal resolution of UED is determined based on the duration of the electron pulses. The beam emittance directly determines the quality of the diffraction image (e.g., the sharpness of the diffraction patterns (DPs) and the diffraction contrast in the acquired images (i.e., spatial resolution)). The pulse duration and emittance of the electron beam are two crucial parameters in UED.

2.1.1 The temporal duration of the electron pulses

The temporal duration of the electron pulses generated from the RF gun, in the absence of time delay due to the response time of the photocathode materials, is

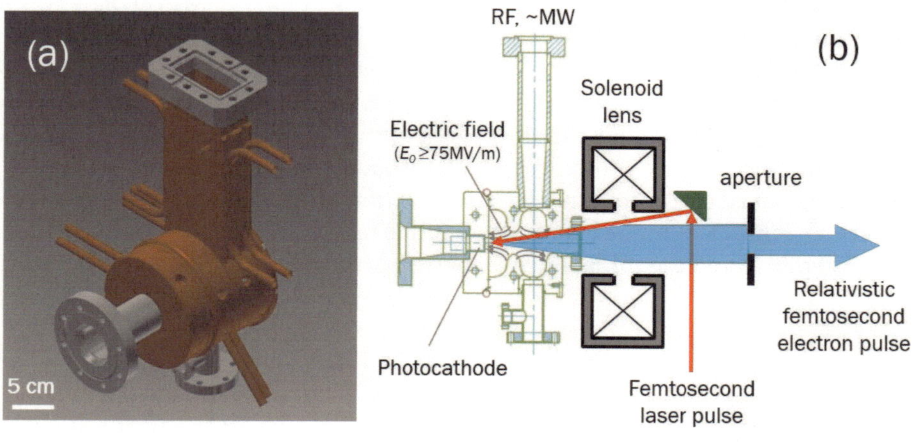

Figure 3.
(a) Photocathode RF gun and (b) schematic for the generation of femtosecond electron pulses in the RF gun.

given by the driving laser pulse duration and temporal electron broadening. The temporal electron broadening can be defined with two components: one derives from the initial energy bandwidth (initial kinetic energy spread, ΔE_{kin}) of electrons emitted from the photocathode, as described in Section 1. Another derives from the space-charge-induced broadening during the propagation from the cathode to the sample. Therefore, the electron pulse duration can be given as

$$\sigma_b = \sqrt{\sigma_{opt}^2 + \tau_{KE}^2 + \tau_{SC}^2}, \tag{2}$$

where σ_b is the pulse duration of electrons, σ_{opt} is the laser pulse duration, and τ_{KE} and τ_{SC} are the temporal broadenings due to the initial kinetic energy spread and space-charge effect, respectively. The first temporal broadening is proportional to $\tau_{KE} \propto \sqrt{\Delta E_{kin}}/E_{acc}$ [12], where E_{acc} is the accelerating electric field (E_{ext} in the dc gun and E_0 for the RF gun). The second temporal broadening is proportional to Q/E^2, where Q and E are the electron charge and total energy of the electron beam, respectively. For the 3-MeV electron pulses at $Q < 1$ pC, τ_{KE} and τ_{sc} in the RF gun are negligible. Therefore, the duration of low-charge electron pulses generated from the RF gun is approximately equal to the driving laser pulse duration, $\sigma_b \approx \sigma_{opt}$.

2.1.2 The total emittance of the electron beam

The total emittance of the electron beam generated from the RF gun [32] is given as:

$$\varepsilon = \sqrt{\varepsilon_{rf}^2 + \varepsilon_{sc}^2 + \varepsilon_{th}^2}, \tag{3}$$

$$\varepsilon_{rf} = 2.73 \times 10^{-11} E_0 f^2 \sigma_x^2 \sigma_b^2, \tag{4}$$

$$\varepsilon_{sc} = 3.76 \times 10^3 \frac{Q}{E_0(2\sigma_x + \sigma_b)}, \tag{5}$$

where ε_{rf} is the normalized RF-induced emittance in root mean square (RMS) in mm-mrad, ε_{sc} is the normalized RMS space-charge-induced emittance in mm-mrad, ε_{th} is the thermal emittance (initial emittance) at the cathode, E_0 is the peak accelerating field in MV/m, f is the RF in MHz, Q is the electron charge of the pulse in nC, and σ_b and σ_x are the RMS pulse duration in ps and RMS transverse beam size in mm, respectively. For example, under the conditions of E_0 = 75 MV/m, σ_b = 100 fs, σ_x = 0.3 mm, and Q = 0.1 pC (10^6 electrons per pulse), we estimate that $\varepsilon_{rf} = 1.5 \times 10^{-5}$ mm-mrad and $\varepsilon_{sc} = 7 \times 10^{-3}$ mm-mrad. This indicates that the RF-induced emittance is negligible and the magnitude of the space-charge-induced emittance at $Q \leq 0.1$ pC is close to the order of nm-rad. Therefore, in this case, the thermal emittance (initial emittance) at the cathode is dominant.

Assuming an isotropic emission into a half sphere in front of the cathode surface, the thermal emittance can be expressed in terms of the RMS incident laser spot size on the cathode σ_r and the initial kinetic energy spread ΔE_{kin} of the photoelectrons:

$$\varepsilon_{th} = \sigma_r \sqrt{\frac{2\Delta E_{kin}}{3m_0c^2}}, \tag{6}$$

where m_0 is the electron mass and σ_r is the laser spot size at the photocathode. For a very fine copper photocathode driven by a 266-nm laser under the conditions of E_0 = 75 MV/m and ϕ_0 = 30°, ΔE_{kin} = 0.42 eV [36]. Therefore, the thermal emittance is estimated as a function of the RMS laser spot size: ε_{th} = 0.74 × σ_r. This indicates that we can reduce the emittance to $\varepsilon_{th} \sim 10$ nm-rad in the RF gun if we can focus the laser spot as σ_r = 10 μm at the photocathode. In this case, the peak brightness of electron pulses B_p can be calculated by

$$B_p = (\beta\gamma)^2 \frac{Q}{\varepsilon^2 \sigma_b}, \qquad (7)$$

where $\beta = v/c$, v is the electron velocity, and γ is the normalized relativistic energy. From a 3-MeV electron pulse with a pulse duration of 100 fs, we can calculate the peak brightness to $B_p = 5 \times 10^{17}$ A/m^2 sr and the beam fluence to $\varphi \sim 2 \times 10^{11}$ m^{-2}. The spatial coherence length can be calculated by [37]:

$$L_c = \frac{h}{m_0 c} \frac{\sigma_x}{\varepsilon}, \qquad (8)$$

where h is Planck's constant. If σ_x = 0.3 mm at the sample, the spatial coherence length is $L_c \sim 10$ nm, which is an ideal value for electron diffraction imaging. It is twice as large or greater than that of current nonrelativistic UED systems [12, 37, 38]. This allows us to detect sharp DPs and acquire good contrast diffraction images in the measurements.

In the presented relativistic UED apparatus, a very fine copper photocathode was used and illuminated by the third-harmonic of a Ti:sapphire laser (266 nm, pulse duration: 90 fs). The pulse energy of the UV light was 5 μJ at maximum. The diameter of the laser spot at the photocathode was 0.1 mm in RMS focused by an optical lens. The injection phase (gun phase) was 30°, which is an optimal condition to minimize the transverse emittance. The electron beam energy was 3 MeV under the 4-MW RF input. The repetition rate of the electron pulses was 10 Hz, which was determined by the repetition rate of the RF pulses.

2.2 UED imaging system

2.2.1 Electron illumination system

The electron illumination system consists of a solenoid magnetic lens, condenser lens, and condenser aperture to control and transfer the electron pulses from the RF gun on the specimen, as shown in **Figure 3**. The solenoid lens with a large beam aperture is used to create a parallel electron beam. The condenser aperture made of a 1-mm-thick molybdenum metal with four pinholes with diameters of 0.3, 0.5, 1, and 2 mm stops the large-divergence electrons to further reduce the emittance, yielding a small illumination convergence angle at the specimen. After the aperture, we use the condenser lens to create a parallel beam or convergent beam on the specimen. The parallel beam is used for selected area diffraction, whereas the convergent beam is used mainly for convergent beam electron diffraction.

The emittance of the electron beam that passed through the aperture with 0.5, 1 and 2 mm diameter pinholes was measured as 0.1, 0.3, and 0.7 mm-mrad [30], respectively. Reducing the emittance increased the RMS brightness in the pulse. The RMS brightness of the transmitted electrons was 2.2, 1.4 and 0.5 × 10^{22} electrons/m^2 sr, and the number of electrons per pulse was ~0.6, 2.5, and 4.4 × 10^7 at the sample with 0.5-, 1-, and 2-mm diameter pinholes, respectively. For the use of the 0.3-mm-diameter pinhole, the number of electrons in the pulse was ~1 × 10^6, and the brightness was estimated to be ≥5 × 10^{22} electrons/m^2 sr. The illumination convergence angle of the electron beam at the sample was α = 26 μrad in the parallel-beam operation mode with the 0.3-mm-diameter condenser aperture, which is discussed in a later section.

The specimen room is located downstream of the condenser lens. The distance from the photocathode to the specimen is 1.2 m. The sample is manipulated by six-axis motorized stages. In a time-resolved experiment, the sample can be pumped by a femtosecond laser pulse, as shown in **Figure 2**. A pump laser with wavelengths of 266, 400, and 800 nm can be selected to meet the requirements of the measured materials. The pulse duration of the pump laser pulses is 90 fs in full width at half maximum (FWHM). The vacuum pressure in the specimen room reaches $\sim 10^{-10}$ Torr. When inserting or changing a new sample, the sample is first installed into a separated vacuum chamber (preparation room) for the sample cleaning.

2.2.2 Imaging system

The diffraction imaging system consists of a diffraction lens (DL) and a projection lens (PL). The diffraction lens focuses the electrons at a back focal plane (BFP), yielding the DPs on the BFP. The projection lens then projects the DPs in the desired magnification onto a viewing screen (scintillator) through a charge-coupled device (CCD) camera, as shown in **Figure 4**. An aperture with a pinhole diameter of 0.5 mm is inserted at the DL center to block scattered electrons and scattered pump laser light. The UED patterns can be observed in two modes: a wide-momentum mode, in which the PL is weak or turned off, and a high-resolution mode, in which the PL magnifies the DPs or images onto the scintillator.

To achieve a high sensitivity to MeV electron detection with a high damage threshold, a Tl-doped CsI columnar crystal scintillator equipped with a fiber optic plate (Hamamatsu Photonics) is used to convert the relativistic-energy DPs or images into optical images. Finally, the optical images are propagated by a thin reflective mirror (at 45°) and detected with an electron-multiplying CCD (EMCCD) with 1024 × 1024 pixels. The effective detection area of the scintillator is 50 × 50 mm^2, whereas the distance from the specimen to the scintillator is 1.6 m. The sensitivity of the whole detection system is 3×10^{-3} counts/electron. The intensity and position of Bragg peaks in the DPs can be monitored and recorded simultaneously using analysis software for studying the structural dynamics.

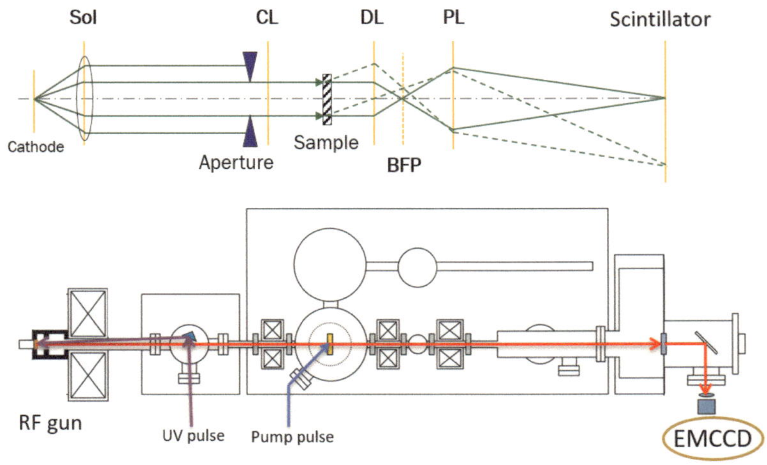

Figure 4.
Schematic of UED imaging with a parallel beam (parallel beam configuration). DL focuses the electrons at a BFP, yielding DPs on BFP. PL then projects the DPs in the desired magnification onto the scintillator. The DPs can be observed in two modes: a wide-momentum mode, in which the PL is weak or turned off, and a high-resolution mode, in which the PL magnifies the DPs or images onto the scintillator.

2.3 Femtosecond laser system

A conventional mode-locked Ti:sapphire femtosecond laser (Spectra-Physics) is used to illuminate the photocathode and to excite the sample. The laser consists of a 80-fs Ti:sapphire laser oscillator (Tsunami, central wavelength: 800 nm) and a regenerative amplifier (Spitfire Ace) that includes a pulse stretcher and compressor. The femtosecond laser oscillator is synchronized to an external 79.3-MHz RF signal with a time-to-lock piezoelectric device, as shown in **Figure 2**. The 79.3-MHz RF signal is generated by dividing the accelerating 2856-MHz RF by 1/36. The time jitter between the laser pulse and RF phase is <100 fs. The laser oscillator output is fed to the regenerative amplifier for pulse stretching, amplification, and compression. The regenerative amplifier is driven by a green laser with a repetition rate of 1 kHz (Empower, wavelength: 532 nm, output: 20 W). The pulse energy of the amplifier output is 3 mJ. The pulse duration is 90 fs in FWHM after the pulse compression.

The amplified femtosecond laser beam is divided into two beams. One is converted to the third-harmonics by a wavelength converter (Tripler) composed of two nonlinear crystals (SHG and THG) and a time plate for pulse delay adjustment. The third-harmonic pulses (UV wavelength, 266 nm; pulse duration, ~90 fs) with a maximum energy of 5 μJ per pulse are focused by an optical lens and then illuminated onto the copper photocathode to generate femtosecond electron pulses. The residual fundamental femtosecond laser (wavelength: 800 nm) is used directly to excite the sample or is converted to second-harmonics (wavelength: 400 nm) or third-harmonics (wavelength: 266 nm) to excite the sample, based on the sample's requirements. The time delay between the pump laser pulse and the probe electron pulse is changed with an optical delay located on the pump laser beam line for time-resolved experiments, as shown in **Figure 2**. The repetition rate of the pump laser pulses is reduced to 10 Hz with two optical choppers, similar to the repetition rate of the electron pulses.

3. UED experiments with relativistic femtosecond electron pulses

3.1 Observations of DPs from crystalline metals, semiconductors, and chemical compounds

In relativistic UED, we measured the DPs of various crystalline materials (e.g., metals, semiconductors, and chemical compounds). The electron pulses generated from the RF gun were collimated by the condenser aperture with the different pinhole diameters of 1, 0.5, and 0.3 mm before the sample was illuminated. The energy and pulse durations of the electron pulses were 3 MeV and ~100 fs, respectively. The following four samples were used:

1. 35-nm-thick single-crystalline silicon (Si) films produced from a 60-μm-thick Si (001) wafer by photolithography and plasma etching

2. 30-nm-thick polycrystalline aluminum foils (Cat. No. S108, EM-Japan)

3. Polycrystals of a thallous chloride chemical compound dispersed on a carbon film pasted on a copper mesh (Cat. No. S110, EM-Japan)

4. Multilayer single-crystalline mica films

All DPs of the four samples were observed under the wide-momentum mode.

Figure 5 shows the DPs of a (001)-orientated single-crystalline Si sample observed both through single-pulse (single-shot) and 10-pulse integrations. A condenser aperture with a 0.3-mm-diameter pinhole was used to collimate the electron beam. The number of electrons in the pulse was $\sim 1 \times 10^6$. In **Figure 5a**, both the lowest Bragg and higher-order peaks clearly appear in the single shot. The excellent quality of the diffraction image is much higher than the pioneering data in the nonrelativistic UED measurement [22]. In **Figure 5b**, entire DPs are clearly visible with 10 pulses integrated. The maximum scattering vector is more than 2 $Å^{-1}$.

The intensity profile of Bragg peaks along (440) and (−4–40) spots of the single-shot image is shown in **Figure 5c**. The RMS width of the zeroth-order spot (000) was 0.015 $Å^{-1}$, indicating an excellent spatial resolution for the MeV diffracted beam. Based on the width of the (000) spot and the measured distance of the diffraction spots from the (000) position, we estimated the RMS illumination convergence angle of the electron beam at the specimen to be α = 26 μrad. By using the 0.3-mm-diameter condenser aperture, we improved both the width and convergence angle from the previous development [20]. The RMS width of both the (220) and (−2–20) diffraction spots, which included the effects of the probe beam energy spread, was identical to that of the (000) spot, indicating a small energy spread in the electron pulse generated by the RF gun.

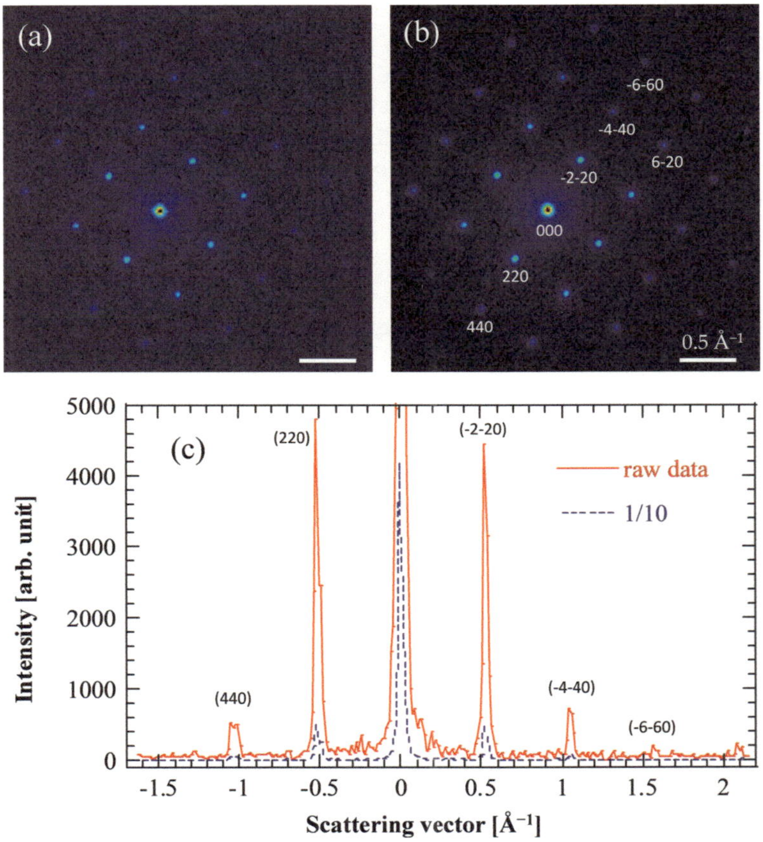

Figure 5.

DPs of a (001)-oriented single-crystalline Si measured with (a) single-shot and (b) 10-pulse integrations. (c) The intensity profile along (440) and (−4–40) spots of the single-shot image. The solid line represents the raw data, and the broken line is given with 1/10 intensities. The energy of the electron pulses is 3 MeV, containing approximately 1×10^6 *electrons per pulse.*

Figure 6 shows transmission electron microscopy (TEM) images and UED patterns of a 30-nm-thick polycrystalline aluminum foil (top images) and polycrystalline thallous chloride (bottom images). The DPs were measured with single-shot and 100-pulse integrations. The energy of the electron pulses was 3 MeV. The condenser aperture with a 1-mm diameter pinhole was used to collimate the electrons. The number of electrons in each pulse was 2.5×10^7. The demonstrations indicate that relativistic UED also enabled the electron diffraction imaging of polycrystalline materials and chemical compounds, and the entire DPs were clearly visible with 100 pulses. These suggest that UED with relativistic femtosecond electron pulses enables the study of ultrafast chemical reactions in chemistry and biology. Moreover, single-shot imaging is also possible for polycrystalline materials. This means that the irreversible processes and reactions in polycrystalline materials and chemical compounds can be observed using relativistic UED.

Figures 7 shows the DPs of a (100)-oriented multilayer single-crystalline mica measured with single-shot, 10-, and 100-pulse integrations [29]. The energy of the electron pulses was 3 MeV. The condenser aperture with a 0.5-mm-diameter pinhole was used to collimate the electrons. The number of electrons in the pulse was 6×10^6. The mica sample was composed of a chemical compound of $KAl_2(AlSi_3)O_{10}(OH)_2$ with a multilayer structure. The thickness of the monolayer was ~10 Å. This is used widely to check the performance of electron diffraction observation in TEM. In general, diffraction imaging of mica is difficult comparing with that of metallic single crystals because of the close diffraction spots. In the relativistic UED measurement as shown in **Figure** 7, both the lowest Bragg and higher-order peaks clearly appear in the observation with 10-pulse integration, and the entire DPs are readily obtained with 100 pulses. Moreover, the possibility of single-shot observation is shown in the measurement.

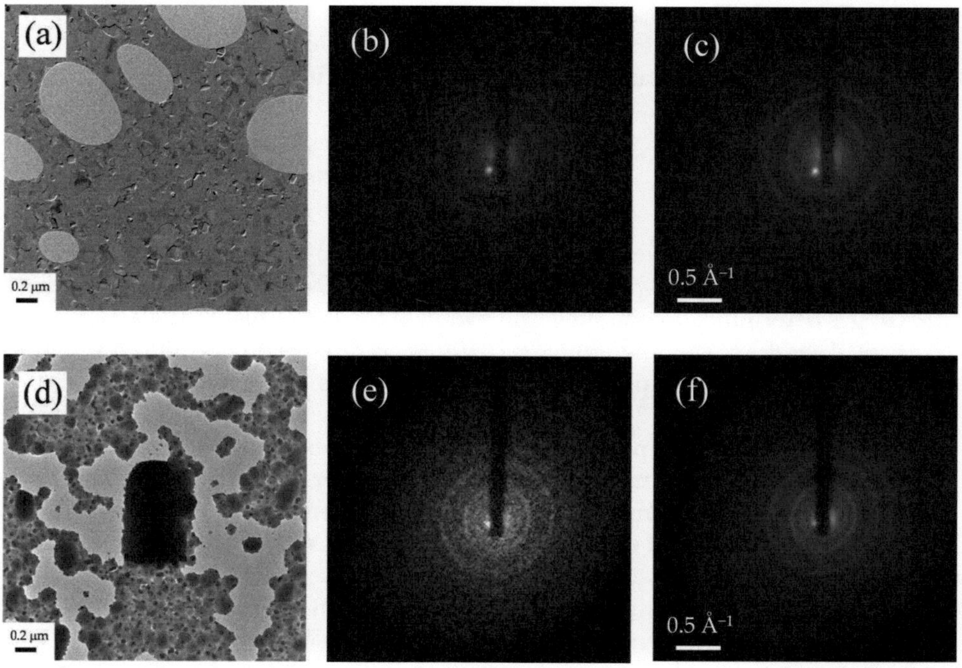

Figure 6.
TEM and UED images. Top: a 30-nm thick polycrystalline Al foil. Bottom: polycrystalline thallous chloride. The images of (a) and (b) are measured by a 200-KV TEM. The DPs of (b) and (e) are measured with single-shot, whereas (c) and (f) are measured with 100-pulse integration. The energy of the electron pulses is 3 MeV, containing 2.5×10^7 *electrons per pulse.*

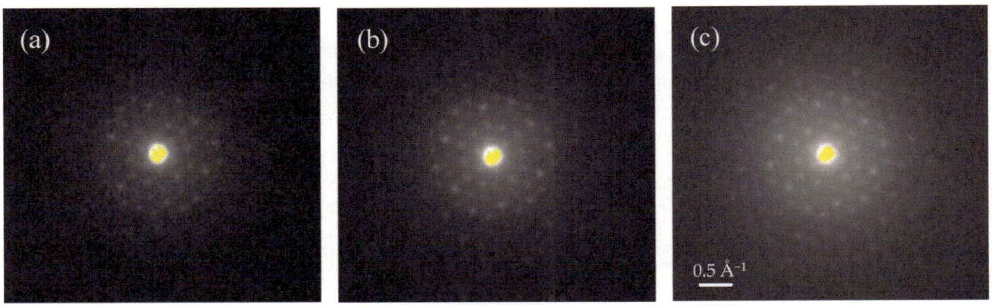

Figure 7.
DPs of a (100)-oriented multilayer single-crystalline mica measured with (a) single-shot, (b) 10-, and (c) 100-pulse integrations [29]. The energy of the electron pulses is 3 MeV, containing 6×10^6 electrons per pulse.

In summary, the experiments indicated that (1) the relativistic UED with femtosecond electron pulses can be applied to electron diffraction imaging of a wide range of materials and (2) relativistic UED enables single-shot observation with femtosecond electron pulses. This means that the ultrafast dynamics of irreversible processes in materials can be measured using relativistic UED.

3.2 Time-resolved measurement for study of ultrafast structural dynamics

A time-resolved measurement technique is used in UED to observe the structural dynamics in a sample. In this case, a femtosecond laser pulse with the desired wavelength based on the sample's requirement is used to excite the sample, whereas the electron pulses are used as a probe to monitor the DPs as a function of the time delay between the pump laser pulse and electron pulse. The time delay in the measurement is adjusted by an optical delay placed on the pump laser beam line, as shown in **Figure 2**. The pulse energy and polarization of the pump laser can be changed to meet the requirements. In the absence of a velocity mismatch between the pump and the probe beams within the sample, the final temporal resolution of UED is expressed as:

$$\Delta t = \sqrt{\sigma_b^2 + \sigma_l^2 + \Delta t_j^2}, \tag{9}$$

where σ_b is the probe electron pulse duration, σ_l is the pump laser pulse duration, and Δt_j represents the time jitter between two pulses. The time jitter in the relativistic UED is determined by the synchronization of the laser pulse to the accelerating RF phase. Therefore, we can define $\Delta t_j < 100$ fs, as discussed in Section 2.3. In the presented time-resolved experiment, both the durations of the pump and probe pulses are ~100 fs in RMS, yielding a total temporal resolution of ~180 fs in RMS.

The intensities of Bragg peaks measured in the UED experiments reflect both the lattice temperature effect (in terms of the Debye-Waller factor) and details of the structural changes. For example, by monitoring the intensities of Bragg peaks as a function of the time delay between the two pulses, we can investigate the phase transformation in materials (e.g., melting of metal excited by laser light). **Figure 8** presents an example of a time-resolved UED experiment to observe laser-induced melting dynamics in a 10-nm-thick (001)-oriented single-crystal gold film [25, 26]. The sample was excited by a 385-nm femtosecond laser (second-harmonic of a turned 770-nm Ti:sapphire laser). The UED patterns were observed by 3-MeV electron pulses. The electron pulses were collimated to a 200-μm diameter by an aperture in the front of the sample. The diameter of the pump laser at the sample was 800 μm, which was much larger than that of the probe electron beam. The

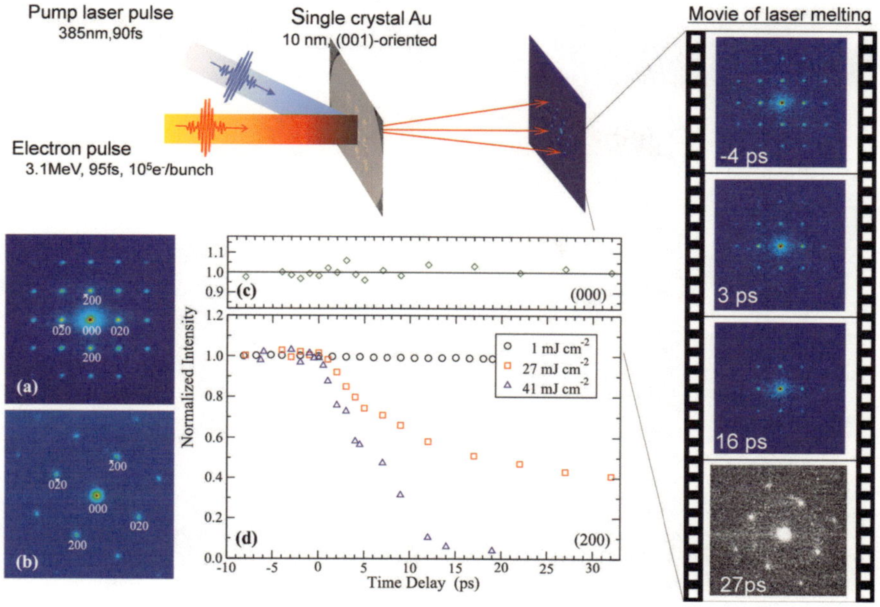

Figure 8.
Schematic and results of a time-resolved UED experiment. The UED patterns (a) and (b) are obtained by wide-momentum and high-resolution modes, respectively. (c) The (000)-order peak intensity was obtained by single-shot at F = 27 mJ/cm². (d) The (200) Bragg peak intensities at F = 1, 27 and 41 mJ/cm² were obtained by averaging over four equivalent (200) Bragg-peak spots [25]. Copyright 2013, with permission from the American Institute of Physics.

temporal resolution of the pump-probe measurement was ~180 fs. **Figure 8d** gives the plot of the (200) Bragg peak intensity as functions of the pump laser fluence (F) and the time delay between the probe and pump pulses. The intensity was normalized with respect to the value measured for each sample prior to laser excitation. At $F = 27$ mJ/cm^2, 40% of the intensity was still detectable after 30 ps, whereas the intensity was nearly reduced to zero within ~15 ps at $F = 45$ mJ/cm^2, indicating a strong dependence on the excitation fluence.

To determine the structural changes in the film reflected in the measured dynamics of Bragg peaks, we applied a hybrid method that combines the two-temperature model with classical molecular dynamics (2 T-MD) [26]. This method has already been used to model the laser-induced melting of nanofilms and nanorods. By comparing the theoretical predictions from 2 T-MD with the UED measurements, we succeeded in revealing the mechanism of the laser-induced melting. **Figure 9** shows a comparison of experimental and theoretical signals at the pump laser fluences of 27 and 41 mJ/cm^2 and the cross sections of the atomistic simulation of melting in the single-crystal gold film at the absorbed fluences (F_{abs}) of 3.0 and 4.5 mJ/cm^2. The model reproduced both the fast decay of Bragg intensity at <5 ps and the slower longer-term behavior. At low fluence, $F = 27$ mJ/cm^2 or $F_{abs} = 3.0$ mJ/cm^2 (**Figure 9a** and **c**), the dynamics exhibited premelting of free surfaces and heterogeneous melting by melt front propagation. The melt fronts slowly propagated toward the center, but the film did not melt entirely, and small regions of crystalline gold remained at 1.2 ns. At high fluence, $F = 41$ mJ/cm^2 or $F_{abs} = 4.5$ mJ/cm^2 (**Figure 8b** and **d**), the sample expanded more rapidly, and the premelting was more pronounced. The average temperature reached the melting temperature at 6 ps, and homogenously distributed seeds of low-density molten phase were subsequently created at 6–12 ps. When the sample reached the limit of crystal stability (after 12 ps), these molten seeds grew and coalesced until the sample melted entirely ~20 ps.

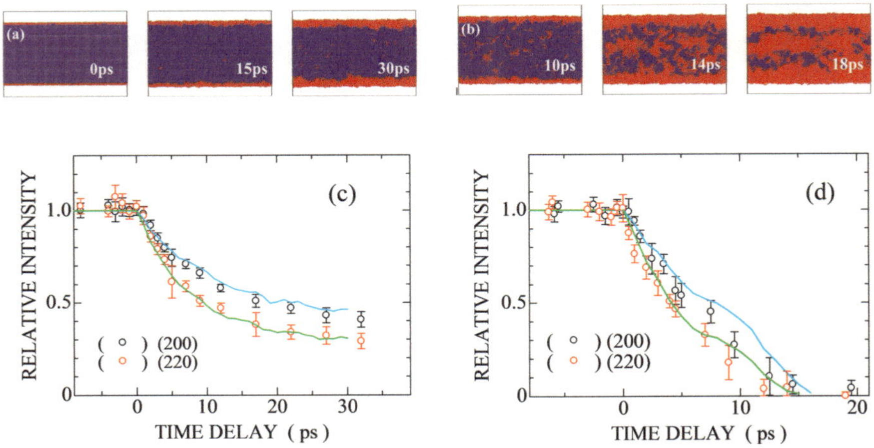

Figure 9.
Cross-sections (top) of the atomistic simulation of meltings in single crystal gold films at (a) F_{abs} = 3.0 and (b) 4.5 mJ/cm^2, and a comparison (bottom) of experimental and theoretical signals as a function of time delay at (c) F = 27 and (d) 41 mJ/cm^2. The dots denote the experimental data, whereas the lines represent the theoretical results. In (a) and (b), all atoms are color-coded according to the measure of a degree of crystallinity given by the (nearest-neighbor averaged) centro-symmetry parameter Φ. Blue atoms ($\Phi < 0.45$) have a local crystalline structure, and red atoms ($\Phi \geq 0.45$) have highly disordered surroundings [25]. Copyright 2013, with permission from the American Institute of Physics.

Recently, other time-resolved observations of ultrafast structural dynamics in semiconductors, organic crystals, and two-dimensional materials have been proposed in relativistic UED. The results showed that relativistic UED has great potential and excellent performance for the study of ultrafast photo-induced phase transitions.

4. Conclusion

In this chapter, we introduced femtosecond diffraction imaging with relativistic electron pulses. We also discussed relativistic femtosecond electron pulse generation using an RF-acceleration-based photoemission gun, a relativistic UED apparatus, and demonstrations of UED measurements with the relativistic femtosecond electron pulses. The electron pulses generated by the RF gun exhibited excellent characteristics, including a low emittance of ≤0.1 mm-mrad, a short pulse duration of 100 fs, and a high number of electrons of $\geq 10^6$ in the pulse at an energy of 3 MeV. The peak brightness was reached at 10^{22} electrons/m^2 sr. These pulses facilitated (1) the acquisition of high-quality DPs with a spatial resolution of 0.015 Å^{-1} and (2) a time-resolved experiment (pump-probe measurement) with an excellent temporal resolution of 180 fs. The most important result was the single-shot observation with relativistic femtosecond electron pulses in a wide range of materials. The findings suggest that relativistic UED is very promising for the study of ultrafast dynamics of irreversible processes and chemical reactions (not discussed in this chapter) in the femtosecond time region.

In addition, the UED system is also very stable because it combines the current state-of-the-art femtosecond mode-locked laser and RF-acceleration and time-synchronization technologies. In our study, the fluctuation of the Bragg peak intensity was <5% in the long term (i.e., 24 hours or more). The relativistic UED apparatus is also very compact.

In the future, an ultralow emittance of ~10 nm-mrad can be expected by focusing the μm-diameter laser spots on the photocathode and collimating the electron beam with a small condenser aperture of 0.1 mm or less. In this case, the spatial coherence length can be improved to $L_c > 10$ nm, which is an ideal value for electron diffraction imaging with the highest spatial resolution. In addition, as discussed in Section 3.2, the atomic-level analytical method of 2 T-MD is very useful to explain the atomic dynamics and underlying mechanisms. Through combination with the 2 T-MD method, relativistic UED can be a powerful tool for femtosecond imaging in materials science, chemistry, and biology.

Acknowledgements

The author acknowledges K. Kan, Y. Yoshida, H. Yasuda, and K. Tanimura of Osaka University and Y. Naruse of the Shiga University of Medical Science for their valuable suggestions and discussions. In addition, the author thanks J. Urakawa, T. Takatomi and N. Terunuma of the High Energy Accelerator Research Organization (KEK) for the design and fabrication of the high-quality RF gun.

This work was supported by a Basic Research (A) (No. 22246127, No. 26246026, and No. 17H01060) of Grant-in-Aid for Scientific Research from MEXT, Japan.

Author details

Jinfeng Yang
The Institute of Scientific and Industrial Research, Osaka University, Osaka, Japan

*Address all correspondence to: yang@sanken.osaka-u.ac.jp

References

[1] Zewail AH. 4D ultrafast electron diffraction, crystallography, and microscopy. Annual Review of Physical Chemistry. 2006;**57**:65-103. Available from: https://www.annualreviews.org/doi/10.1146/annurev.physchem.57.032905.104748

[2] Siwick BJ, Dwyer JR, Jordan RE, Dwayne Miller RJ. An atomic-level view of melting using femtosecond electron diffraction. Science. 2003;**302**: 1382-1385. DOI: 10.1126/science.1090052

[3] King WE et al. Ultrafast electron microscopy in materials science, biology, and chemistry. Journal of Applied Physics. 2005;**97**:111101. DOI: 10.1063/1.1927699

[4] Rood AP, Milledge J. Combined flash-photolysis and gas-phase electron-diffraction studies of small molecules. Journal of the Chemical Society, Faraday Transactions II. 1984;**80**: 1145-1153. Available from: https://pubs.rsc.org/en/content/articlepdf/1984/f2/f29848001145

[5] Ewbank JD et al. Instrumentation for gas electron diffraction employing a pulsed electron beam synchronous with photoexcitation. The Review of Scientific Instruments. 1992;**63**: 3352-3358. DOI: 10.1063/1.1142552

[6] Lobastov VA et al. Instrumentation for time-resolved electron diffraction spanning the time domain from microseconds to picoseconds. The Review of Scientific Instruments. 1998; **69**:2633-2643. DOI: 10.1063/1.1148991

[7] Mourou GA, Williamson S. Picosecond electron diffraction. Applied Physics Letters. 1982;**41**:44-45. DOI: 10.1063/1.93316

[8] Williamson S, Mourou G, Li JCM. Time-resolved, laser-induced phase transformation in aluminium. Physical Review Letters. 1984;**52**:2364-2367. DOI: 10.1103/PhysRevLett.52.2364

[9] Ihee H et al. Direct imaging of transient molecular structures with ultrafast diffraction. Science. 2001;**291**: 458-462. DOI: 10.1126/science.291.5503.458

[10] Williamson JC et al. Clocking transient chemical changes by ultrafast electron diffraction. Nature. 1997;**386**: 159-162. DOI: 10.1038/386159a0

[11] Cao J et al. Femtosecond electron diffraction for direct measurement of ultrafast atomic motions. Applied Physics Letters. 2003;**83**:1044-1046. DOI: 10.1063/1.1593831

[12] Aidelsburger M, Kirchner FO, Krausz F, Baum P. Single-electron pulses for ultrafast diffraction. PNAS. 2010;**107**:19714-19719. DOI: 10.1073/pnas.1010165107

[13] Siwick BJ, Dwyer JR, Jordan RE, Dwayne Miller RJ. Ultrafast electron optics: Propagation dynamics of femtosecond electron packets. Journal of Applied Physics. 2002;**92**:1643-1648. DOI: 10.1063/1.1487437

[14] Siwick BJ, Dwyer JR, Jordan RE, Dwayne Miller RJ. Ultrafast electron optics: Propagation dynamics of femtosecond electron packets. Chemical Physics. 2004;**299**:285-305. DOI: 10.1016/j.chemphys.2003.11.040

[15] Sciaini G, Dwayne Miller RJ. Femtosecond electron diffraction: Heralding the era of atomically resolved dynamics. Reports on Progress in Physics. 2011;**74**:096101. DOI: 10.1088/0034-4885/74/9/096101

[16] Hassan MT. Attomicroscopy: From femtosecond to attosecond electron microscopy. Journal of Physics B:

Atomic, Molecular and Optical Physics. 2018;**51**:032005. DOI: 10.1088/1361-6455/aaa183

[17] Hastings JB, Rudakov FM, Dowell DH, Schmerge JF, Cardoza JD, Castro JM, et al. Ultrafast time-resolved electron diffraction with megavolt electron beam. Applied Physics Letters. 2006;**89**:184109. DOI: 10.1063/1.2372697

[18] Li RK, Tang CX, Du YC, Huang WH, Du Q, Shi JR, et al. Experimental demonstration of high quality MeV ultrafast electron diffraction. The Review of Scientific Instruments. 2009;**80**:083303. DOI: 10.1063/1.3194047

[19] Musumeci P, Moody JT, Scoby CM. Relativistic electron diffraction at the UCLA Pegasus photoinjector laboratory. Ultramicroscopy. 2008;**108**:1450-1453. DOI: 10.1016/j.ultramic.2008.03.011

[20] Murooka Y, Naruse N, Sakakihara S, Ishimaru M, Yang J, Tanimura K. Transmission-electron diffraction by MeV electron pulses. Applied Physics Letters. 2011;**98**:251903. DOI: 10.1063/1.3602314

[21] Zhu P, Zhu Y, Hidaka Y, Wu L, Cao J, Berger H, et al. Femtosecond time-resolved MeV electron diffraction. New Journal of Physics. 2015;**17**:063004. DOI: 10.1088/1367-2630/17/6/063004

[22] Harb M, Peng W, Sciaini G, Hebeisen CT, Ernstorfer R, Eriksson MA, et al. Excitation of longitudinal and transverse coherent acoustic phonons in nanometer free-standing films of (001) Si. Physical Review B. 2009;**79**:094301. DOI: 10.1103/PhysRevB.79.094301

[23] Shen X, Li RK, Lundstrom U, Lane TJ, Reid AH, Weathersby SP, et al. Femtosecond mega-electron-volt electron microdiffraction. Ultramicroscopy. 2018;**184**:172-176. DOI: 10.1016/j.ultramic.2017.08.019

[24] Fu F, Liu S, Zhu P, Xiang D, Zhang J, Cao J. High quality single shot ultrafast MeV electron diffraction from a photocathode radio-frequency gun. The Review of Scientific Instruments. 2014;**85**:083701. DOI: 10.1063/1.4892135

[25] Giret Y, Naruse N, Daraszewicz SL, Murooka Y, Yang J, Duffy DM, et al. Determination of transient atomic structure of laser-excited materials from time-resolved diffraction data. Applied Physics Letters. 2013;**103**:253107. DOI: 10.1063/1.4847695

[26] Daraszewicz SL, Giret Y, Naruse N, Murooka Y, Yang J, Duffy DM, et al. Structural dynamics of laser-irradiated gold nanofilms. Physical Review B. 2013;**88**:184101. DOI: 10.1103/PhysRevB.88.184101

[27] Yang J, Kan K, Naruse N, Yoshida Y, Tanimura K, Urakawa J. 100-femtosecond MeV electron source for ultrafast electron diffraction. Radiation Physics and Chemistry. 2009;**78**:1106-1111. DOI: 10.1016/j.radphyschem.2009.05.009

[28] Yang J, Yoshida Y, Shibata H. Femtosecond time-resolved electron microscopy. Electronics and Communications in Japan. 2015;**98**:50-57. DOI: 10.1002/ecj.11763

[29] Yang J. Ultrafast electron microscopy using relativistic-energy femtosecond electron pulses. Microscopy. 2015;**50**:156-159. Available from: http://microscopy.or.jp/jsm/wp-content/uploads/publication/kenbikyo/50_3/50_3e04jy.html

[30] Yang J, Yoshida Y, Yasuda H. Ultrafast electron microscopy with relativistic femtosecond electron pulses. Microscopy. 2018;**67**:291-295. DOI: 10.1093/jmicro/dfy032

[31] Yang J, Yoshida Y. Relativistic ultrafast electron microscopy: Single-shot diffraction imaging with femtosecond electron pulses. Advances in Condensed Matter Physics. 2019; **2019**:9739241. DOI: 10.1155/2019/9739241

[32] Arita M, Sakaguchi N. Electron microscopy: Novel microscopy trends. In: Yang J, editor. Ultrafast Electron Microscopy with Relativistic Femtosecond Electron Pulses. InTechOpen; 2019. DOI: 10.1093/jmicro/dfy032

[33] Kan K, Yang J, Kondoh T, Yoshida Y. Development of femtosecond photocathode RF gun. Nuclear Instruments and Methods in Physics Research Section A. 2011;**659**: 44-48. DOI: 10.1016/j.nima.2011.08.016

[34] Terunuma N, Murata A, Fukuda M, Hirano K, Kamiya Y, Kii T, et al. Improvement of an S-band RF gun with a Cs_2Te photocathode for the KEK-ATF. Nuclear Instruments and Methods in Physics Research Section A. 2010;**613**: 1-8. DOI: 10.1016/j.nima.2009.10.151

[35] Yang J, Sakai F, Yanagida T, Yorozu M, Okada Y, Takasago K, et al. Low-emittance electron-beam generation with laser pulse shaping in photocathode radio-frequency gun. Journal of Applied Physics. 2002;**92**: 1608-1612. DOI: 10.1063/1.1487457

[36] Yang J, Kan K, Kondoh T, Murooka Y, Naruse N, Yoshida Y, et al. An ultrashort-bunch electron RF gun. Journal of the Vacuum Society of Japan. 2012;**55**:42-49. DOI: 10.3131/jvsj2.55.42

[37] Van Oudheusden T, de Jong EF, van der Geer SB. Op't root WPEM, Luiten OJ. Electron source concept for single-shot sub-100 fs electron diffraction in the 100 keV range. Journal of Applied Physics. 2007;**102**:093501. DOI: 10.1063/1.2801027

[38] Gahlmann A, Park AT, Zewail AH. Ultrashort electron pulses for diffraction, crystallography and microscopy: Theoretical and experimental resolutions. Physical Chemistry Chemical Physics. 2008;**10**: 2894-2909. DOI: 10.1039/b802136h

Chapter 3

Femtosecond Pulse Radiolysis

Jinfeng Yang, Koichi Kan, Masao Gohdo and Yoichi Yoshida

Abstract

Ultrafast pulse radiolysis with a short-pulsed electron beam and a short-pulsed analyzing light is a powerful time-resolved spectroscopic technique to study the kinetics and reactions of short-lived intermediate species or precursors in radiation chemistry and biology. In this chapter, first, we give an overview of historical developments of ultrafast pulse radiolysis. Then, we describe a femtosecond pulse radiolysis instrument, including the generation of femtosecond electron pulses by a photocathode radio frequency (rf) gun-based linear electron accelerator, the synchronization of femtosecond analyzing laser with the electron pulses, the transient absorption measurement with double-pulse technique, and the observations of the formation processes and ultrafast reactions of hydrated electrons in water. Finally, two innovative techniques, which enable to improve the time resolution in next pulse radiolysis development, are presented.

Keywords: pulse radiolysis, femtosecond electron pulse, femtosecond laser, short-lived intermediate species, precursors, hydrated electron, ultrafast reaction, radiation chemistry

1. Introduction

1.1 Primary processes in radiation chemistry

Primary processes or ultrafast reactions in radiation chemistry and biology are occurred in the ionization of molecules with radiation beams, for example, gamma- or X-rays, electrons, ions, and other high-energy particles. The ionization of molecules produces a positive radical ion and an electron with an initial kinetic energy. The electron will be thermalized by interactions with molecules, and then surrounded by solvent molecules to form solvated electron state in polar materials like water [1–3], or reacted with the positive radical ion by geminate recombination in nonpolar materials like alkenes [4–10].

For example, in water, the primary processes at the current state of knowledge in the radiation chemistry [11, 12] can be summarized as:

$$H_2O \rightsquigarrow H_2O^{\bullet+} + e^-, H_2O \rightsquigarrow H_2O^* \quad (1)$$

$$H_2O^{\bullet+} + H_2O \rightarrow OH^\bullet + H_3O^+ \quad (2)$$

$$H_2O^* \rightarrow OH^\bullet + H^\bullet \quad (3)$$

$$e^- + nH_2O \rightarrow e^-_{hyd} \quad (4)$$

where Eq. (1) represents the ionization and electronic excitation of water molecules on a attosecond timescale ($\sim 10^{-16}$ s), Eq. (2) is the ion-molecular reaction of positive radical ion $H_2O^{\bullet+}$ in $\sim 10^{-14}$ s, Eq. (3) is the dissociation of electronic excited state H_2O^* in $\sim 10^{-13}$ s, and Eq. (4) is the formation process of solvated electron (called also hydrated electron in water, denoted by e^-_{hyd}) in $\sim 10^{-13}$ s.

To study such ultrafast kinetics and reactions, two techniques have been developed: photolysis and pulse radiolysis. The photolysis is mainly used in photochemistry, while pulse radiolysis is widely used in radiation chemistry. In photolysis, short-lived intermediate species or precursors are produced by photoionization with high-energy photons or multiphoton excitation. The species and their reactions are measured by time-resolved absorption spectroscopy with analyzing laser light. The first observation of femtosecond solvation dynamics of hydrated electrons in water was achieved in 1987 by Migus et al. by photolysis with a femtosecond laser [1]. In pulse radiolysis, short-pulsed radiation, that is, pulsed X-rays, electrons, or ions, ionizes molecules. The short-lived intermediate species and primary processes are observed by transient absorption spectroscopy with an analyzing light. The first pulse radiolysis was developed in 1962 by Hart and Boag [13] using 1.8-MeV and 3-μs-long electron pulses generated by an electron accelerator. They succeeded firstly in the direct spectroscopic observation of the solvated electrons in aqueous solutions. The experimental results indicated that pulse radiolysis is a very promising technique to determine short-lived intermediate species or precursors and observe their kinetics or reactions.

1.2 Historical developments of ultrafast pulse radiolysis

Progress of particle accelerator technology pays a significant role in the ultrafast pulse radiolysis development. The first picosecond pulse radiolysis was developed at the University of Toronto in the late 1960s [14] using the fine structure of ~30-ps electron pulse train with the duration of 30 ns generated by S-band (2856 MHz) linear electron accelerator (or linac). Cherenkov light emitted by the picosecond electron pulses was used as analyzing light to detect the transient absorption, as shown in **Figure 1**. However, in this pulse radiolysis, the sample was irradiated every 350 ps (one period of S-band microwave) by the electron pulse train. Reactions or kinetics with time constant of >350 ps overlapped in the measurement. To circumvent this problem, several picosecond pulse radiolysis facilities using picosecond single-pulse electron accelerators with a sub-harmonic bunching technique were constructed in 1970s and 1980s. Tabata et al. at the University of Tokyo developed a 20-ps pulse radiolysis with two parallel linacs [15–17]. One linac was used to generate a picosecond single-pulse electron beam to irradiate the sample.

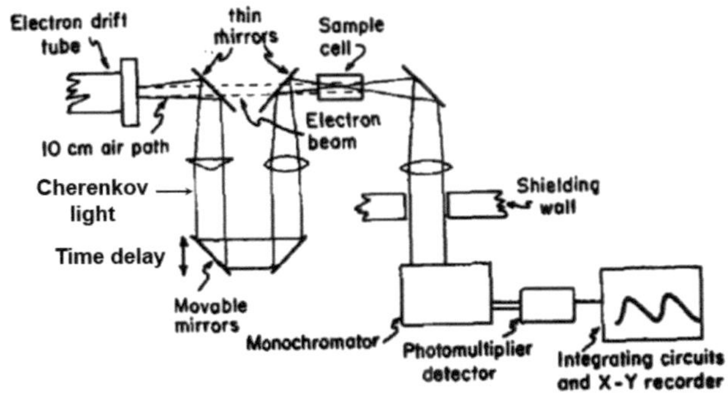

Figure 1.
First picosecond pulse radiolysis with time resolution of 23 ps developed at the University of Toronto in the late 1960s [14].

Another linac was used to produce picosecond single-pulse Cherenkov light. In 1990s, a picosecond pulse radiolysis system using the laser diode instead of the Cherenkov light was developed [18]. The system enabled to measure the transient absorption from the visible to near-infrared region. In the late 1990s, a femtosecond laser was used for the analyzing light to cover the wide wavelength range from ultraviolet to infrared [19]. However, in spite of many innovative techniques that have been developed, the time resolution had remained at ~20 ps since the 1970s.

A remarkable progress in ultrafast pulse radiolysis has been made since 2000. A sub-picosecond pulse radiolysis was developed at Osaka University using a sub-picosecond single-pulse electron beam and a femtosecond laser light [20]. The sub-picosecond single-pulse electron beam was generated by a 28-MeV L-band (1300 MHz) electron linac and a magnetic pulse compressor. A synchronized Ti: sapphire laser with pulse width of 60 fs was used as analyzing light source. Moreover, a time jitter compensation technique was developed by directly detecting the time interval between the electron pulse and the analyzing laser light with a femtosecond streak camera. Finally, the time resolution of 2 ps was achieved.

In the 2000s, a new type of ultrafast pulse radiolysis using an advanced accelerator technology of laser-triggered photocathode radio frequency (rf) electron gun was developed at the Brookhaven National Laboratory, at the University of Tokyo, at Sumitomo Heavy Industries, at the University of Paris-Sud, at Waseda University, and at Osaka University. The photocathode rf gun consists of two to four rf cavities driven with high rf power to accelerate the electrons emitted from the photocathode to an energy region of 4~9 MeV. The rf gun enables to generate a picosecond single-pulsed electron beam with electron charge of nano-coulombs using picosecond laser excitation on the photocathode. The new type of pulse radiolysis exhibits two crucial advantages: (1) the instrument is very compact without the need of the sub-harmonic bunching system and (2) the electron pulses produced by the rf gun are time-synchronized with laser pulses, resulting in a low time jitter between the electron pulse and the analyzing laser light. **Table 1** gives the ultrafast pulse radiolysis facilities using the photocathode rf guns.

In 2010, the first femtosecond pulse radiolysis with time resolution of 240 fs was successfully developed at Osaka University using a femtosecond single-pulsed electron beam generated by a 32-MeV photocathode rf gun linac and a magnetic pulse compressor [26–29]. The femtosecond pulse radiolysis was used successfully to observe the femtosecond solvation dynamics of hydrated electrons in water.

Facility	Instrument	Electron pulse characteristics	Refs.
Brookhaven National Laboratory	3.5-cell rf gun, ~100-fs laser	≥7 ps, 6~8 nC, 8.7 MeV	[21]
University of Tokyo	1.6-cell rf gun, booster linac, 100-fs laser	7 ps, 0.65 nC, 18 MeV, time jitter: 2.1 ps	[22]
Sumitomo Heavy Industries	1.6-cell rf gun, 15-ps laser	20 ps, 1 nC, 1.75 MeV	[23]
University of Paris-Sud	1.5-cell rf gun, booster linac, 120-fs laser	<5 ps, <2 nC, 10 ps, 5 nC, 4~9 MeV	[24]
Waseda University	1.6-cell rf gun, 4.8-ps white light	6 ps, 1 nC, 4.5 MeV	[25]
Osaka University	1.6-cell rf gun, booster linac, 107-fs laser	98~400 fs, 0.1~1 nC, 32 MeV, time jitter: 61 fs	[26–29]

Table 1.
The ultrafast pulse radiolysis facilities using photocathode rf guns.

It paves the way to observe the short-lived intermediate species and primary processes in radiation chemistry on femtosecond timescale.

Recently, a technique of laser wakefield electron acceleration instead of the traditional accelerators was applied in ultrafast pulse radiolysis at the Argonne National Laboratory [30]. However, the time resolution is still limited to picosecond because of the electron pulse duration, the time jitter, and the effect due to group velocity mismatch (GVM) between the electron beam and the light in the sample.

In order to reduce furthermore the time jitter between the electron pulse and the analyzing light pulse, a technique of a "double-decker electron beam accelerator" was proposed in pulse radiolysis at Osaka University [31, 32]. The entirely synchronized double-decker electron beams with a time interval of 1.4 ns were generated in the photocathode rf gun with 2-ps laser beams, then accelerated to 32 MeV with a booster linac, and finally compressed into femtosecond with a magnetic compressor. One of the double-decker electron beams was used to irradiate the sample, while another beam was converted to the analyzing light, for example, Cherenkov light or terahertz (THz) light, resulting in the generation of entirely synchronized femtosecond-pulsed electron beam and femtosecond analyzing light. Moreover, in order to circumvent the effect of GVM, an innovative technique of "equivalent velocity spectroscopy (EVS)" was developed for pulse radiolysis at Osaka University [33]. The method of ultrafast pulse radiolysis was reviewed in a book entitled "Recent trends in radiation chemistry" edited by Wishart and Rao in 2010 [34].

In this chapter, we describe the details of femtosecond pulse radiolysis, including the generation of femtosecond electron pulses, the synchronization of femtosecond analyzing laser with the electron pulses, double-pulse technique for the transient absorption measurement, and the observations of formation processes and ultrafast reactions of hydrated electrons in water. Finally, two innovative techniques of EVS and double-decker electron beam accelerator will be introduced.

2. Femtosecond pulse radiolysis

Ultrafast pulse radiolysis is widely constructed with a stroboscopic method (a pump-and-probe method), as shown in **Figure 2**. The electron pulse is used as a radiation pulse, while the analyzing light pulse (laser or Cherenkov light) is used to measure the time evolution of transient absorption of short-lived intermediate species by changing the time interval between the electron pulse and the analyzing light pulse. In the stroboscopic method, the time resolution is determined by the electron pulse duration, the analyzing light pulse duration, and the time jitter between the electron pulse and the analyzing light pulse, as described in Section 3.1.

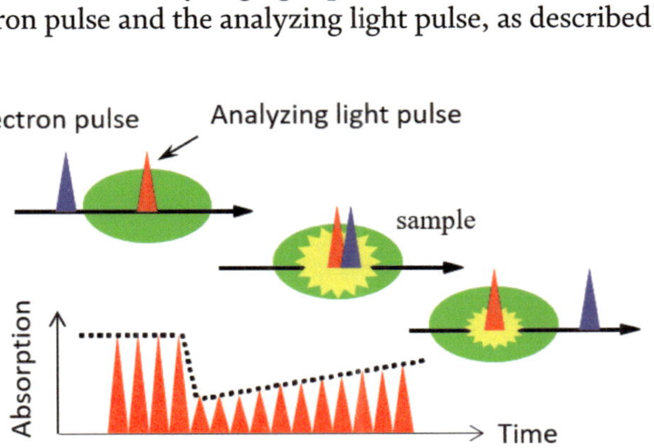

Figure 2.
Schematic of stroboscopic method (pump-and-probe method) in ultrafast pulse radiolysis.

The progress of the ultrafast laser technology allows us to use a femtosecond laser as an analyzing light source in pulse radiolysis. However, in order to achieve a femtosecond time resolution, three significant techniques are indispensable:

1. generating a femtosecond-pulsed electron beam,

2. synchronizing the electron pulse with the analyzing light pulse in femtosecond, and

3. detecting the transient absorption with a good signal-to-noise (S/N) ratio in a thin sample cell to circumvent the effect of GVM.

In this section, we focus on the generation of femtosecond electron pulses with a laser photocathode electron gun accelerator, ultrafast pulse radiolysis with the femtosecond-pulsed electron beam and a femtosecond laser, and a double-pulse detection technique to observe the transient absorption in a thin sample cell.

2.1 1.6-cell photocathode rf gun

Femtosecond pulse radiolysis at Osaka University was constructed with a 1.6-cell S-band (2.856 GHz) photocathode rf gun, a booster linac, a magnetic pulse compressor, and an analyzing femtosecond Ti:sapphire laser, as shown in **Figure 3**.

The photocathode rf gun used in femtosecond pulse radiolysis contains two cavities: a half cell and a full cell. The length of the half-cell cavity was equal to 0.6

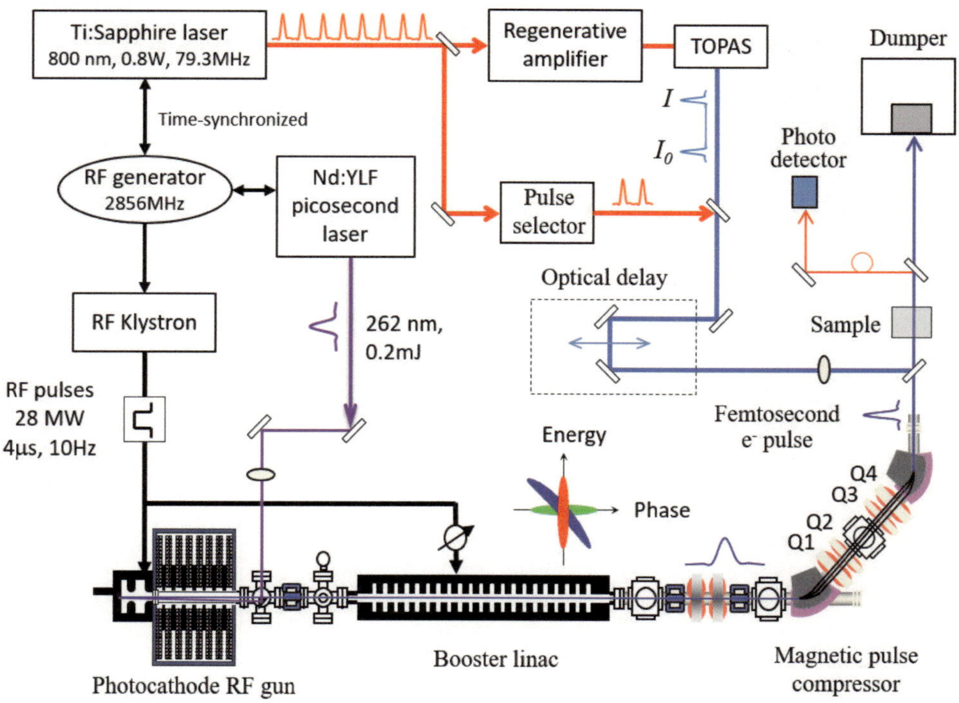

Figure 3.
Schematic of femtosecond pulse radiolysis apparatus at Osaka University, containing a photocathode rf gun, a booster linac, a magnetic pulse compressor, and an analyzing femtosecond Ti:sapphire laser. The rf gun was driven by a picosecond Nd:YLF laser. A time resolution of 240 fs due to the electron beam and the analyzing laser has been achieved [26–29].

times $\lambda/2$ to minimize the beam emittance, while $\lambda = 104.96$ mm is the wavelength of 2.856-GHz rf. The cavities were operated with 10-MW peak rf power to produce a high-peak electric field of ∼110 MV/m on photocathode and the cavities [35–39]. A copper cathode was used and illuminated by the fourth harmonic of Nd:YLF picosecond laser [262 nm, pulse duration: 5 ps in full width at half maximum (FWHM)], as described in Section 2.4. The incident angle of the laser light was approximately 2° along the electron beam direction using a mirror placed in vacuum. The laser injection phase (gun phase) was adjusted by changing the phase of the reference 79.3-MHz rf signal with a low-power phase shifter. A solenoid magnet was mounted at the exit of the rf gun to reduce the emittance growth due to the space charge effect during the propagation. The maximum magnetic field was 3 kG.

The characteristics of the electron beam generated by the rf gun are shown in **Table 2**. **Figure 4** gives the transverse emittance, pulse duration, and relative energy spread in root-mean-square (RMS) as a function of the laser injection phase. The rf gun enables to generate a low-emittance and low-energy-spread electron beam at the laser injection phase of <40°. Moreover, the pulse compression inside the rf gun due to the rf effect occurred at <40°, yielding a electron pulse duration less than that of the incident laser. However, the electron charge was decreased at a low laser injection phase, that is, 1.8 nC at 80° was decreased to 0.2 nC at 10°. In the pulse radiolysis experiment, the gun phase was fixed to 30° in the experiments.

Electron energy	4 MeV
Electron charge	1 nC/pulse
Pulse duration	1.8 ± 0.2 ps
Transverse emittance	3.2 ± 0.2 mm-mrad at 1 nC
Relative energy spread	0.05%
Repetition rate of pulses	10 Hz

Table 2.
The characteristics of picosecond electron beam generated by the rf gun at the laser injection phase of 30° [35].

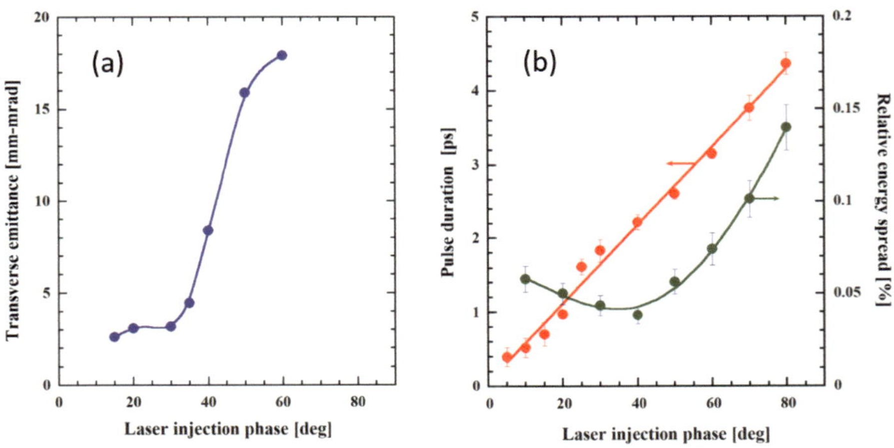

Figure 4.
The transverse emittance, pulse duration, and relative energy spread as a function of the laser injection phase [35]. The electron charge was 1.8 nC at 80° and was decreased to 0.2 nC at 10°.

2.2 Booster linear accelerator

A 2-m-long S-band traveling-wave booster linac was mounted downstream of the rf gun and the solenoid magnet at a distance of 1.2 m from the photocathode. The booster linac was driven by rf pulses with peak power of 25 MW. The duration of the rf pulses was 4 μs. The repetition rate of the pulses was 10 Hz. The booster linac accelerates the electron pulses and makes an optimum energy-phase correction for pulse compression, where the electrons in the front of pulse has more energy than the electrons in the back of pulse, as shown as the blue pulse in **Figure 3**. The rf phase of the booster linac was adjusted by a high-power phase shifter mounted in the 25-MW rf line. The energy of the electrons was 32 MeV after the booster linac.

2.3 Magnetic pulse compressor

The magnetic pulse compressor consists of two 45°-bending magnets and four quadrupole magnets. To obtain an ultrashort electron pulse, all magnets were carefully installed with the minimum lattice error to reduce the aberrations in the phase space distribution. The outside two quadrupole magnets (Q1 and Q4) have equal magnetic fields, while the inside two quadrupole magnets (Q3 and Q4) have equal magnetic fields. They provide the necessary path length dependence on the electron energy. The dispersion function is symmetric on the mid-plane of the compressor. When the picosecond electron pulse with an optimum energy-phase correction (blue pulse in **Figure 3**) passes the magnetic pulse compressor, the pulse is compressed into femtosecond by rotating the pulse in longitudinal phase space, as shown as red pulse in **Figure 3**. However, higher order momentum dispersion in the compressor, especially the second-order dispersion, causes a nonlinear deformation of the longitudinal phase space, yielding the increase of the final pulse duration. To minimize the nonlinear effects, a nonlinear energy correlation of the electron pulse (blue pulse in **Figure 3**) was introduced before the compression by re-phasing the booster linac to >90° with a curvature of the rf waveform. Finally, the correlation offsets the effects of the nonlinear path length in the magnetic compression to obtain a short electron pulse [28].

Figure 5(a) gives the compressed pulse duration and the relative energy spread as a function of the rf phase of the booster linac at 1 nC [28]. The data show that the shortest pulse duration of 400 ± 58 fs RMS was obtained at 94°. At this phase, the booster linac accelerates the electrons with a nonlinear energy-phase correlation. The pulse duration was decreased by decreasing the electron charge, that is, a 98-fs-long electron pulse at 0.17 nC.

2.4 Laser system

The laser system contains an all solid-state LD-pumped Nd:YLF picosecond laser (Time-Bandwidth) and a Ti:sapphire femtosecond laser (Spectra-Physics). The Nd:YLF picosecond laser was used to generate a picosecond electron beam in the rf gun. The laser consists of a 79.3-MHz passive mode-locked Nd:YLF laser oscillator, a regenerative amplifier, and a wavelength converter. The 79.3-MHz laser pulses were phase-locked with a reference 79.3-MHz rf signal, which was generated by dividing the accelerating 2.856-GHz rf by 1/36, by dynamically adjusting the laser cavity length with a semiconductor saturable absorber mirror (SESAM) and a timing stabilizer. A single laser pulse of the oscillator was captured and amplified to the pulse energy up to 1 mJ in the regenerative amplifier. The repetition rate of the amplified pulses was 10 Hz. The amplified laser pulses were converted to ultraviolet

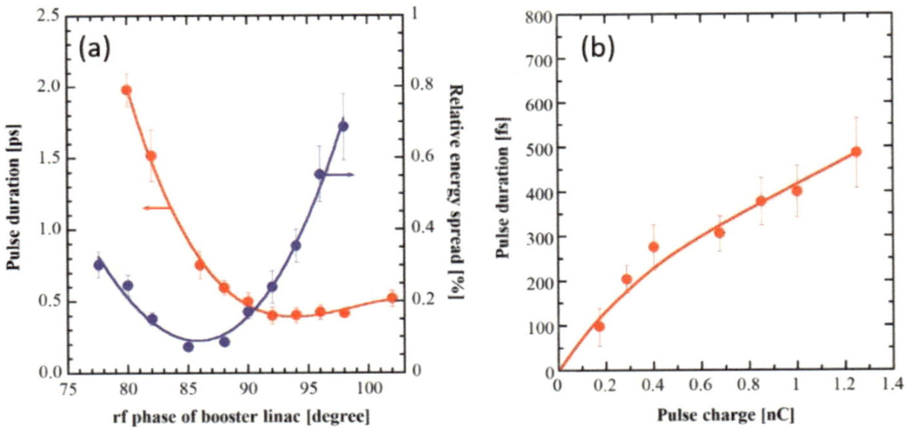

Figure 5.
(a) The compressed pulse duration and the relative energy spread as a function of the rf phase of booster linac at 1 nC and (b) the compressed pulse duration as a function of electron charge at the rf phase of 94° [28]. Copyright 2006, with permission from Elsevier.

pulses (UV: 262 nm) using two nonlinear crystals and guided to the photocathode in the rf gun. The pulse duration of UV light was 5 ps FWHM. The maximum pulse energy was 300 μJ.

The mode-locked Ti:sapphire femtosecond laser (Spectra-Physics) was used as analyzing light source in pulse radiolysis. The laser consists of a 79.3-MHz femtosecond Ti:sapphire laser oscillator (Tsunami, central wavelength: 800 nm), a regenerative amplifier (Spitifire), and a tunable optical parametric amplifier with frequency mixer stages (TOPAS-Prime and NirUVis), as shown in **Figure 6(a)**. The 79.3-MHz laser pulses were synchronized to the external 79.3-MHz RF signal with a time-to-lock piezoelectric device. The time jitter between the laser pulse and 79.3-MHz rf phase was 61 fs RMS [26], being approximately equal to the time jitter between the electron pulse and the analyzing laser pulse. The laser oscillator output was fed to the regenerative amplifier for pulse stretching, amplification, and compression. The regenerative amplifier was driven by a green laser with a repetition rate of 1 kHz (Empower, wavelength: 532 nm, output: 15 W). The pulse energy of the amplifier output was 0.8 mJ. The pulse duration was ~100 fs after the pulse compression. The amplified femtosecond laser beam was inputted to the optical

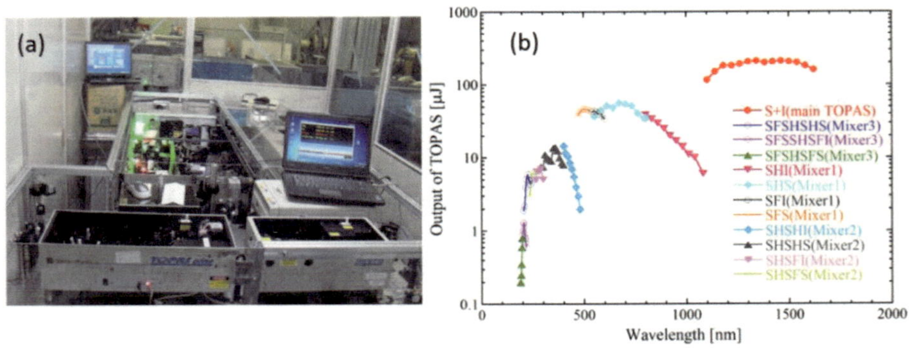

Figure 6.
(a) Photo of TOPAS-Prime and NirUVis and (b) the pulse energy of analyzing light as a function of wavelength, which facilitates pulse radiolysis [40]. Copyright 2017, with permission from Japan Radioisotope Association.

parametric amplifier to extend the tuning range into ultraviolet (UV), visible (VIS), or infrared (IR). **Figure 6(b)** shows the pulse energies of outputs of TOPAS-Prime and NirUVis [40]. An analyzing light with a wide wavelength range of 190~1600 nm facilitates pulse radiolysis based on the sample's requirements. The time delay between the electron pulse and the analyzing laser pulse was adjusted by changing the arrival time of the laser pulse at the sample with an optical delay.

2.5 Double-pulse technique: transient absorption measurement

Progress in the photodiode technology allows us to easily detect the transient absorption of intermediate species over a wide wavelength range from ultraviolet to infrared, that is, using silicon photodetectors in wavelength range from 200 to 1000 nm (PDA10A, Thorlabs) and InGaAs photodetectors in wavelength range from 800 to 1700 nm (PDA10C, Thorlabs). However, the development of transient absorption detection technique with a good S/N ratio is very significant in ultrafast pulse radiolysis, especially for the use of low-charge femtosecond electron pulses and a thin sample cell to circumvent the effect of GVM.

For this purpose, we developed a double-pulse measurement technique to reduce the fluctuation of the laser intensity caused by long-term drift and mechanical vibration of optics. **Figure 7** shows the concept of the double-pulse measurement. Two analyzing laser pulses with a time interval of Δt are incident on the sample. The front laser pulse is used as "reference pulse" before the electron irradiation, while the back laser pulse is used as a "signal pulse" to measure the absorption after the electron irradiation. The optical density is thus obtained by

$$OD = \log \frac{I_0}{I} \tag{5}$$

where I_0 and I are the signals of the reference and signal pulses detected by a photodiode, respectively. The time interval between two pulses was Δt = 1 ms using the outputs of the optical parametric amplifier.

The output of the Ti:sapphire laser oscillator (Tsunami) with the wavelength of 800 nm was also utilized in pulse radiolysis [29]. The continuous pulses of 79.3 MHz were guided to a pulse selector (Spectra-Physics, 3980), which was constructed with an Acousto-Optic Modulator (AOM) crystal driven by the 79.3-MHz rf pulses. The selector extracted a pulse train containing several pulses by adjusting the rf pulse duration and phase. Finally, two stable pulses of the pulse train with a same

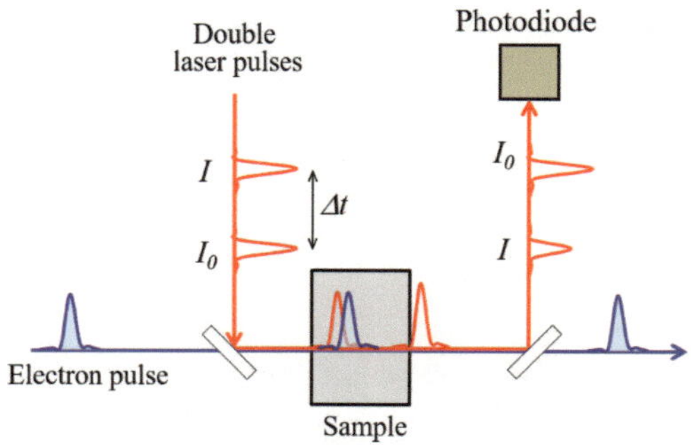

Figure 7.
Schematic of double-pulse measurement in pulse radiolysis to reduce the fluctuation of the laser intensity caused by long-term drift and mechanical vibration of optics.

intensity were used as the double-pulse measurement. The time interval between two pulses was Δt = 12.6 ns. The previous studies [26, 27] indicate that this technique enables to detect a small optical density of 0.001 with an acceptable S/N ratio in a thin sample cell with optical length of 0.18 mm.

3. Pulse radiolysis experiments

3.1 Observations of solvation dynamics of hydrated electrons in water

In order to reduce the degradation of time resolution due to GVM, a thin sample cell with an optical length of 0.18 mm was used in the observation of hydrated electrons [26]. The water sample was deionized and Ar-saturated before the measurement. The double pulses with time interval of 12.6 ns generated by the 79.3-MHz Ti:sapphire laser oscillator and the pulse selector were used as analyzing light. The wavelength was 800 nm with a bandwidth of 12.5 nm FWHM. The pulse energy was approximately 10 nJ. The duration of the laser pulses was 107 fs RMS. The intensities of the reference pulse and the signal pulse were detected by a silicon photodiode and a digital phosphor oscilloscope. A femtosecond electron beam with pulse duration of 201 fs RMS at 0.4 nC generated by the accelerator system was used to irradiate the sample. The time jitter between the laser pulse and the electron pulse was measured to 61 fs RMS.

Figure 8 shows the first observation of ultrafast transient absorption kinetics of hydrated and pre-hydrated electrons in water pulse radiolysis. The transient absorption kinetics of pre-hydrated electrons were observed in a water sample cell with an optical length of 1 mm at the wavelength of 1300 nm. The kinetics observed in **Figure 8(a)** are in agreement with the formation process of hydrated electrons in the multiphoton ionization with two-state model, in which the hydrated electron is formed via a short-lived precursor (called "pre-hydrated electron," denoted by e^-_{pre}) as

$$e^-_{th} \xrightarrow{\tau_1} e^-_{pre} \xrightarrow{\tau_2} e^-_{hyd} \tag{6}$$

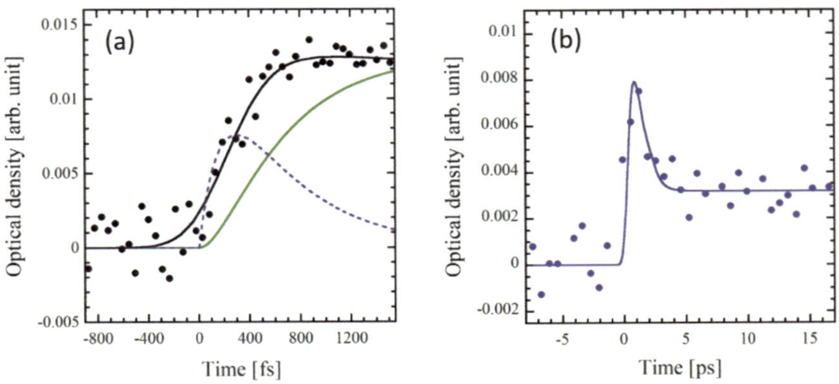

Figure 8.
(a) Femtosecond transient absorption kinetics of hydrated electrons observed in a thin water sample with an optical length of 0.18 mm at the wavelength of 800 nm, and (b) picosecond transient absorption kinetics of pre-hydrated electrons observed in a water sample cell with an optical length of 1 mm at the wavelength of 1300 nm. The blue and green lines in ***Figure 8(a)*** *show the decay of pre-hydrated electrons and the formation of hydrated electrons with a time constant of 550 fs. The black line represents the kinetics of reaction (6) with* τ_1 *= 180 fs and* τ_2 *= 550 ± 50 fs. The data at ≤ 3 ps in* ***Figure 8(b)*** *are mainly contributed by the pre-hydrated electrons, while the data at > 3 ps represent the kinetics of hydrated electrons with a long time decay. Copyright 2017, with permission from Japan Radioisotope Association.*

where e^{-}_{th} represents the thermalized electron. When we fit the data in **Figure 8(a)** with the reaction (6), we found that the hydrated electrons are formed by the decay of pre-hydrated electrons with a time constant of τ_2 = 550 ± 50 fs and the pre-hydrated electrons are formed within τ_1 = 110~200 fs after the electron irradiation. The obtained formation time of hydrated electrons in water pulse radiolysis is in agreement with that of τ_1 = 540 fs obtained in multiphoton ionization studies [2, 3].

The time resolution of pulse radiolysis based on the stroboscopic method can be estimated by two components: one is the time resolution due to the electron beam and the analyzing laser light, and another is the time resolution due to GVM in the sample. The time resolution due to the electron beam and the laser light can be estimated by

$$\delta t_b = \sqrt{\sigma_e^2 + \sigma_l^2 + \sigma_j^2} \tag{7}$$

where σ_e = 201 fs RMS is the electron pulse duration, σ_l = 107 fs RMS is the analyzing laser pulse duration, and σ_j = 61 fs RMS is the time jitter between the two pulses, yielding δt_b = 240 fs.

The time resolution due to GVM in the sample with the experimental setup as shown in **Figure 7** can be calculated by

$$g(L) = L\left(\frac{n}{c} - \frac{1}{v}\right) \tag{8}$$

where L = 0.18 mm is the optical length of the sample, n is the refractive index of the sample, n = 1.33 for water sample, c is the velocity of light in vacuum, and v is the velocity of the electrons ($v \cong c$ for the 32-MeV electrons), yielding $g(L)$ = 198 fs. Therefore, the total resolution of pulse radiation was $\Delta\tau = \delta t_b + g(L)$ = 438 fs, which is available to observe the formation process of hydrated electrons with a time constant of 550 ± 50 fs in water as shown in **Figure 8**.

3.2 Observations of ultrafast spur reactions in water pulse radiolysis

Although the formation time of hydrated electrons in pulse radiolysis is similar to that in the multiphoton ionization, the thermalization distance (length of initial distribution) of electrons in pulse radiolysis is longer than that in the photoionization because the electrons produced by radiation ionization have a high initial kinetic energy. It causes some difference of spur reactions (primary processes) in the early time, containing the reactions of hydrated electrons with H_3O^+ cation and $OH^{\bullet}$ radical pair:

$$e^-_{hyd} + H_3O^+ \rightarrow H_2O + H \tag{9}$$

$$e^-_{hyd} + OH\cdot \rightarrow OH^- \tag{10}$$

Figure 9 shows the time evolution of hydrated electrons observed in water pulse radiolysis at 800 nm [40] and obtained in photolysis with a photoexcitation energy of 8.3 eV [41]. The data show that a small magnitude (~10%) of the hydrated electrons decreases in the first 40 ps due to the spur reactions (reactions (9) and (10)) in pulse radiolysis, while 38% of hydrated electrons reacted with H_3O^+ cation and $OH^{\bullet}$ radical pair in photolysis with a photoexcitation energy of 8.3 eV. We fit the data by a numerical calculation of the survival probability of the hydrated electrons in the reactions (9) and (10) with a Gaussian initial distribution of hydrated electrons in water as in the photolysis study [41]:

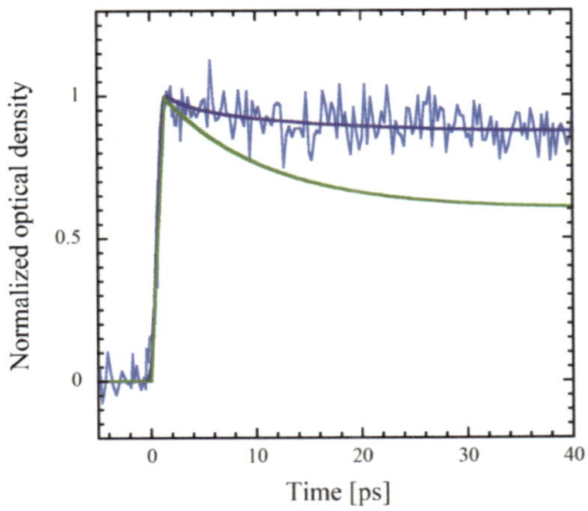

Figure 9.
Time evolution of hydrated electrons observed in water pulse radiolysis at 800 nm (blue lines) and in photolysis with a photoexcitation energy of 8.3 eV (green line) [41]. Copyright 2017, with permission from Japan Radioisotope Association.

$$f(r_0) = \frac{1}{\sqrt{8\pi^3\sigma^6}} \exp\left(-\frac{r_0^2}{2\sigma^2}\right), \quad \text{and} \quad \langle r_0 \rangle = \sigma\sqrt{8/\pi} \tag{11}$$

where $\langle r_0 \rangle$ is the distance of the hydrated electron distribution (the average thermalization distance). We found $\langle r_0 \rangle$ = 7~9 nm in pulse radiolysis, which is much longer than $\langle r_0 \rangle$ = 0.9 nm in the photolysis with a photoexcitation energy of 8.3 eV. The long distance in water pulse radiolysis causes a small magnitude of the hydrated electrons reacting with H_3O^+ cation and $OH^{\bullet}$ radical pair in the first 40 ps.

The raw data in **Figure 9** give the ratio of optical density at 1.5 and 20ps, $OD_{1.5ps}/OD_{20ps}$ = 1.12. Using a hydrated electron yield (G-value) of 4.1 per 100 eV obtained at 20ps in previous water pulse radiolysis studies [21, 22], we then obtained G = 4.6 ± 0.3 at 1.5ps, which is as good as the initial yield of hydrated electrons in water pulse radiolysis because the thermalized electrons are fully hydrated at 1.5ps. The obtained G-value is in agreement with the initial yield of 4.8 estimated in scavenger studies [42].

3.3 Equivalent velocity spectroscopy

The degradation of time resolution due to GVM in the sample can be reduced using a thin sample cell. However, the use of thin sample cell leads to a small absorption signal (optical intensity) and the degradation of S/N ratio in the measurement. On the other hand, due to the space charge effect, the ultrashort electron pulses are realized only at low electron charge, that is, sub-20-fs electron pulses at 2.1 pC [39]. Therefore, the increase of both the time resolution and the absorption signal is a great challenge in pulse radiolysis.

To circumvent the GVM problem and increase the absorption signal, we proposed an innovative technique of equivalent velocity spectroscopy (EVS) in pulse radiolysis [33]. The approach of EVS is shown in **Figure 10**. The electron pulse and the analyzing laser pulse are incident on the sample with an angle (θ) determined by the refractive index (n) of the sample as cosθ = $1/n$. If we rotate the electron pulse with an angle (φ) and φ = θ, both the rotated electron pulse and the analyzing light pulse precisely overlap at every point in the sample, resulting in the entire circumvention of GVM between the electron beam and the light in the sample.

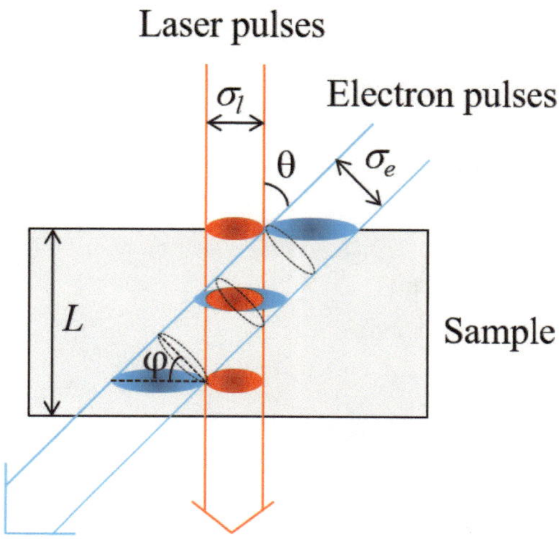

Figure 10.
Schematic of EVS in pulse radiolysis with rotated electron pulses and analyzing laser pulses [33]. Copyright 2009, with permission from Elsevier.

In EVS, the time resolution due to GVM in the sample can be thus calculated by

$$g(L) = \sigma_e\left(\frac{\sin\varphi}{v_e\cos\varphi}+\frac{\cos\theta}{v_e\sin\theta}-\frac{1}{v_l\sin\theta}\right)+\sigma_l\left(\frac{1}{v_e\sin\theta}-\frac{\cos\theta}{v_l\sin\theta}\right)\text{ for } L > \frac{\sigma_l\cos\theta}{\sin\theta}+\frac{\sigma_e}{\sin\theta}, \quad (12)$$

where σ_e and σ_l are the sizes of the electron beam and the analyzing light respectively,φ is the rotation angle of the electron pulse, v_e and v_l are the velocities of the electrons and the analyzing light in the sample respectively, with $\approx c$ for the 32-MeV electrons and $v_l = c/n$,. If $\varphi = \theta$ in Eq. (12), $g(L) = 0$. It means that the resolution limitation due to GVM in the sample is thus fully removed. Moreover, a long optical length (L) can be used in EVS.

The technique of the electron pulse rotation is very significant in EVS. In order to rotate the electron pulse, first, we adjusted the accelerating rf phase in the booster linac to increase the relative energy spread in the electron pulse, shown as the red pulse at the exit of the linac in **Figure 11(a)**. Then, we transported the electron pulse through the magnetic pulse compressor. The electron pulse was thus rotated along the propagation direction, shown as the red pulse at the exit of the compressor in **Figure 11(a)**. The rotation angle increased by increasing the accelerating rf phase in the booster linac. **Figure 11(b)** shows the 2D images of the electron pulses with different rotation angle at exit of the pulse compressor measured by a femtosecond streak camera at the accelerating rf phase of ϕ_{rf} = 94°, 99°, and 103° [33]. The electron charge was 1.8 nC per pulse. At ϕ_{rf} = 94°, an optimal energy-phase correlation in the electron pulse was produced for the bunch compressor, as described in Section 2.3. The electron pulse was compressed into the shortest pulse duration of 2 ps FWHM or 0.8 ps RMS, but the electron pulse was not rotated. At ϕ_{rf} = 99°, the rotation angle of the electron pulse was φ = 41°, which is a required angle for water sample. However, the pulse duration was increased to 5 ps. It was caused mainly by the large relative energy spread in the magnetic pulse compressor. At ϕ_{rf} = 103°, φ= 59°, and the pulse duration was increased to 6 ps.

In the demonstration of EVS, we measured the transient absorption kinetics of hydrated electrons in water using rotated electron pulses (φ = 41°) and unrotated electron pulses (φ = 0°). A sample cell with an optical length of 10 mm was used. The electron beam and the analyzing light were incident on the sample with the angle of θ = 41°. The electron charge was 1.8 nC per pulse. The duration of the

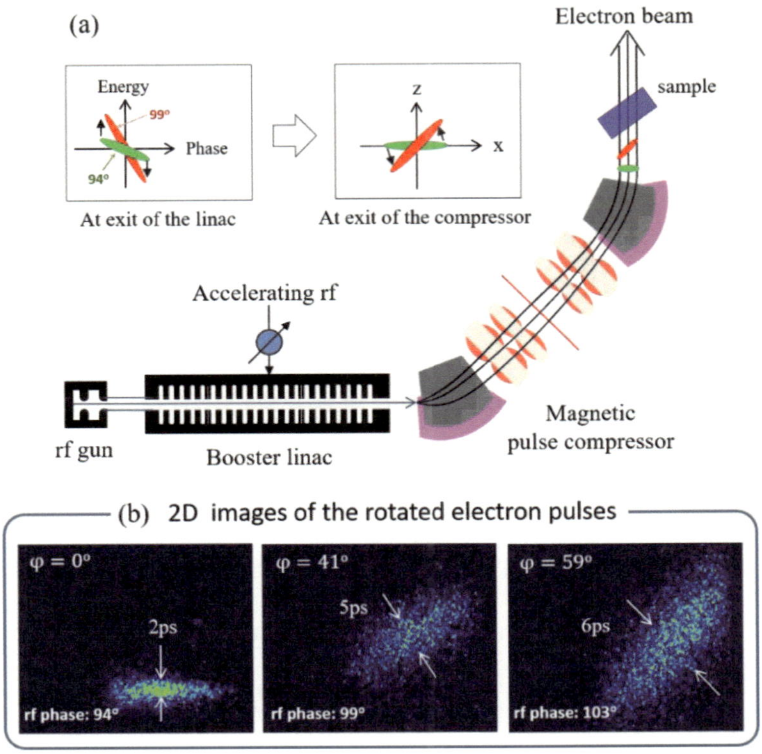

Figure 11.
(a) Schematic of electron pulse rotation using the booster linac and magnetic pulse compressor. (b) 2D images of electron pulses with different rotation angle at exit of the pulse compressor measured by a femtosecond streak camera at the accelerating rf phase of ϕ_{rf} *= 94°, 99°, and 103° [33]. Copyright 2009, with permission from Elsevier.*

electron pulses was 2 ps at φ = 0° and 5 ps at φ = 41°, as shown in **Figure 11(b)**. The double-pulse measurement with time interval of 12.6 ns, output of the 79.3-MHz Ti:sapphire laser oscillator, was used. The duration of the laser pulses was 107 fs RMS. The sizes of the electron beam and the analyzing laser light used in EVS were σ_l = 2 mm and σ_e = 3 mm, respectively.

Figure 12 gives the transient absorption kinetics of hydrated electrons in water measured by EVS. The rise time of the signals was 11.4 ps with the unrotated electron pulses and was reduced to 6.4 ps with the rotated electron pulses. A higher optical density was obtained with the rotated electron pulses. Under the experimental conditions, we calculated the time resolution due to GVM in the sample to g(L) = 8.7 ps using the unrotated electron pulses (φ = 0°) and g(L) = 0 ps using the rotated electron pulses (φ = 41°), according to Eq. (12). On the other hand, the time resolution due to the electron beam and the analyzing light was calculated in Eq. (7) to be $\delta t_b \cong 2$ ps for the unrotated electron pulses and $\delta t_b \cong 5$ ps for the rotated electron pulses. The total time resolution in EVS pulse radiolysis was estimated to be $\Delta\tau$ = 10.7 ps using unrotated electron pulses and $\Delta\tau$ = 5 ps using unrotated electron pulses, which are in a good agreement with the obtained rise time in the measurements using both the unrotated and rotated electron pulses.

The results indicate that (1) EVS is a very promising technique to circumvent the degradation of the time resolution due to GVM in the sample, and (2) higher optical densities can be obtained in EVS with the rotated electron pulses because two pulses overlap at every point in the sample and a thick sample can be used. However, the technique of the electron pulse rotation without the increase of the pulse duration remains to be developed.

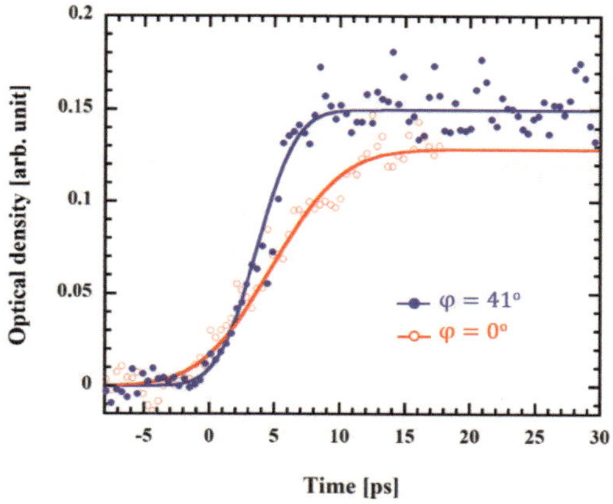

Figure 12.
Transient absorption kinetics of hydrated electrons in water measured in EVS pulse radiolysis with the rotated electron pulses (blue solid dots) and unrotated electron pulses (red circles) [33]. The wavelength of the analyzing laser light was 800 nm. Copyright 2009, with permission from Elsevier.

3.4 Double-decker electron beam accelerator

Recently, progress of the advanced particle accelerator and ultrashort laser technologies enables the generation of attosecond electron and laser pulses. When using such ultrashort electron and laser pulses in pulse radiolysis, the time jitter between the two pulses becomes a serious problem. Here, we introduce a technology of double-decker electron beam accelerator to generate both entirely synchronized electron pulse and analyzing light pulse for pulse radiolysis.

Figure 13 shows the concept of the double-decker electron beam accelerator for pulse radiolysis [31, 32]. The output of Nd:YLF picosecond laser was divided by a beam splitter to produce 2-ps laser beams with different positions in the vertical direction. The 2-ps laser beams were illuminated the photocathode in the rf gun. An optical delay was mounted in the up beam line to time delay the up beam. Two electron beams generated in the rf gun were accelerated in the booster linac up to 32 MeV and were then compressed into femtosecond with the magnetic pulse compressor.

Figure 13(b) shows the images of two electron beams with up and down positions at the exit of the compressor, which are called double-decker electron beams. In the experiment, the spot sizes of two laser beams on the photocathode were approximately 1 mm. The distance between the two beams was 2 mm in the vertical direction. The signals of the double-decker electron beams measured by a current transformer are shown in **Figure 13(c)**. The electron charges of double-decker beams were 0.47 nC per pulse (up-beam) and 0.65 nC per pulse (down beam). The different charges were due to the different pulse energies of two laser beams produced in the beam splitter. The time interval of the double-decker electron pulses was 1.4 ns, which is equal to four periods (4×0.35 ns) of the accelerating 2856-MHz rf. The double-decker electron beams have a low emittance of 2.5 ± 0.6 mm-mrad for the up beam and 3.6 ± 0.7 mm-mrad for the down beam, and a low relative energy spread of $0.14 \pm 0.03\%$ for both the beams. These enable to compress the double-decker pulses into femtosecond with the magnetic pulse compressor. **Figure 14** shows the temporal distributions of the double-decker electron pulses after the pulse compression measured by a femtosecond streak camera [31]. The pulse duration was obtained to 430 ± 25 fs FWHM for the up beam and 510 ± 20 fs FWHM for the down beam. The difference in the pulse duration for the double pulses was due to the different pulse charge. The two beams

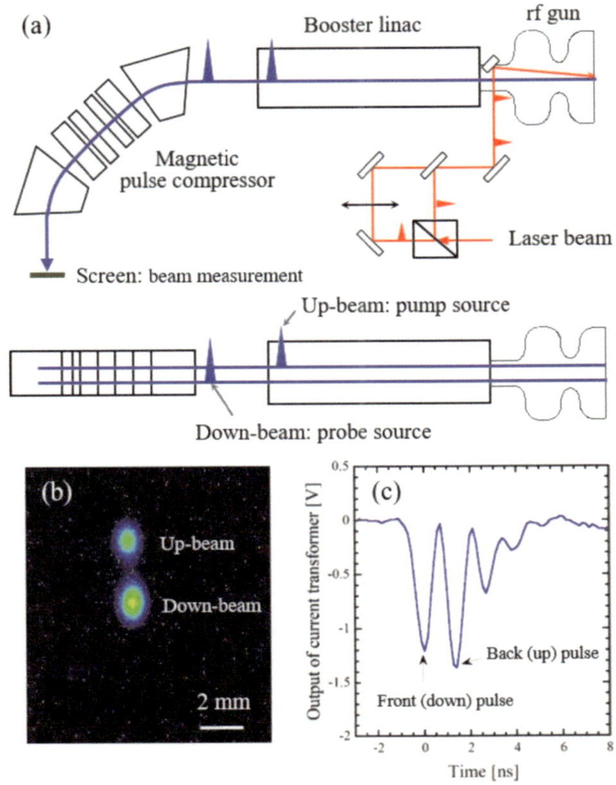

Figure 13.
(a) Top view and side view of double-decker electron beam accelerator, (b) images and (c) signals of double-decker electron beams [31]. Copyright 2009, with permission from American Institute of Physics.

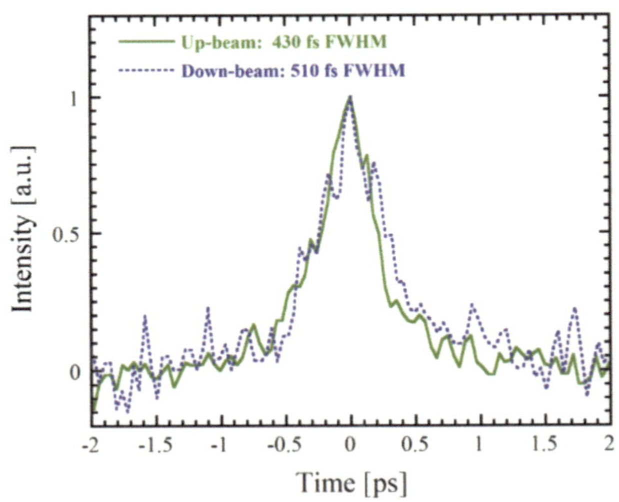

Figure 14.
Temporal distributions of the double-decker electron pulses after the pulse compression measured by a femtosecond streak camera [31]. The pulse duration of the up beam was 430 ± 25 fs FWHM (green solid line), while the pulse duration of the down beam was 510 ± 20 fs FWHM (blue dashed line). Copyright 2009, with permission from American Institute of Physics.

were generated by a laser and accelerated within an accelerating rf pulse. The theoretical estimation indicates that the time jitter in double-decker electron pulses can be reduced to attosecond by using a stable accelerating rf.

In pulse radiolysis, the front electron pulse was converted to the femtosecond analyzing light with Cherenkov radiation, while the back electron pulse irradiated the sample. A band-pass filter was used to select a necessary wavelength for measurement. An optical delay was used to change the time delay between the Cherenkov light pulse and the irradiating electron pulse. In the demonstration, we succeeded to observe the time evolution and spectrum of transient absorption of hydrated electrons in water [32].

4. Conclusion

In this chapter, we introduced a femtosecond pulse radiolysis instrument, including the generation of femtosecond electron pulses, synchronization of femtosecond analyzing light with the electron pulses, a technique of double-pulse measurement to improve S/N ratio, and the first observation of femtosecond formation processes of hydrated electrons in water. A time resolution of 240 fs due to the electron beam and the analyzing laser has been achieved [26–29]. The femtosecond pulse radiolysis enables the observation of the transient absorption in a thin sample cell with an optical length of 0.18 mm, and paves the way to observe the short-lived intermediate species and primary processes in radiation chemistry and biology on femtosecond timescale.

Two innovative techniques namely “equivalent velocity spectroscopy (EVS)” and “double-decker electron beam accelerator” were presented for next pulse radiolysis development. EVS enables to circumvent entirety the effect of group velocity mismatch (GVM) between the electron beam and the analyzing light in the sample. The double-decker electron beam technique can be expected to reduce the time jitter between the electron pulse and the light pulse to attosecond. Moreover, the double-decker electron beam accelerator enables to generate both ultrashort pulse electron beam and analyzing light. Of course, the electron pulse rotation without the increase of pulse duration and some other problems remain to be solved; however, a combination of the EVS technology with the double-decker electron beams enables the development of ultrafast pulse radiolysis with sub-femtosecond or attosecond time resolution to reveal the entire primary processes beginning ionization in radiation chemistry and biology.

Author details

Jinfeng Yang*, Koichi Kan, Masao Gohdo and Yoichi Yoshida
The Institute of Scientific and Industrial Research, Osaka University, Osaka, Japan

*Address all correspondence to: yang@sanken.osaka-u.ac.jp

References

[1] Migus A, Gauduel Y, Martin JL, Antonetti A. Excess electrons in liquid water: First evidence of a prehydrated state with femtosecond lifetime. Physical Review Letters. 1987;**58**:1559-1562. DOI: 10.1103/PhysRevLett.58.1559

[2] Long FH, Lu H, Eisenthal KB. Femtosecond studies of the presolvated electron: An excited state of the solvated electron? Physical Review Letters. 1990; **64**:1469-1472. DOI: 10.1103/PhysRevLett.64.1469

[3] Laenen R, Roth T, Laubereau A. Novel precursors of solvated electrons in water: Evidence for a charge transfer process. Physical Review Letters. 2000;**85**:50-53. DOI: 10.1103/PhysRevLett.85.50

[4] Hong KM, Noolandi J. Solution of the Smoluchowski equation with a coulomb potential. I. General results. The Journal of Chemical Physics. 1987;**68**:5163-5171. DOI: 10.1063/1.435636

[5] Tagawa S, Washio M, Kobayashi H, Katsumura Y, Tabata Y. Picosecond pulse radiolysis studies on geminate ion recombination in saturated hydrocarbon. Radiation Physics and Chemistry. 1983;**21**:45-52. DOI: 10.1016/0146-5724(83)90128-0

[6] Tagawa S, Hayashi N, Yoshida Y, Washio M, Tabata Y. Pulse radiolysis studies on liquid alkanes and related polymers. Radiation Physics and Chemistry. 1989;**34**:503-511. DOI: 10.1016/1359-0197(89)90053-2

[7] Yoshida Y, Tagawa S, Kobayashi H, Tabata Y. Study of geminate ion recombination in a solute-solvent system by using picosecond pulse radiolysis. Radiation Physics and Chemistry. 1987;**30**:83-87. DOI: 10.1016/1359-0197(87)90088-9

[8] Yang J, Kondoh T, Norizawa K, Nagaishi R, Taguchi M, Takahashi K, et al. Picosecond pulse radiolysis: Dynamics of solvated electrons in ionic liquid and geminate ion recombination in liquid alkanes. Radiation Physics and Chemistry. 2008;**77**:1233-1238. DOI: 10.1016/j.radphyschem.2008.05.031

[9] Kondoh T, Yang J, Norizawa K, Kan K, Yoshida Y. Femtosecond pulse radiolysis study on geminate ion recombination in n-dodecane. Radiation Physics and Chemistry. 2011;**80**: 286-290. DOI: 10.1016/j.radphyschem.2010.07.049

[10] Kondoh T, Yang J, Norizawa K, Kan K, Kozawa T, Ogata A, et al. Femtosecond pulse radiolysis study of geminate ion recombination in biphenyl–dodecane solution. Radiation Physics and Chemistry. 2013;**84**:30-34. DOI: 10.1016/j.radphyschem.2012.06.051

[11] Allen AO. The Radiation Chemistry of Water and Aqueous Solutions. Princeton, N.J.: Van Nostrand; 1961

[12] Rodgers MAJ. Farhataziz, Radiation Chemistry: Principles and Applications. VCH: Weinheim; 1987

[13] Hart EJ, Boag JW. Absorption spectrum of the hydrated electron in water and in aqueous solutions. Journal of the American Chemical Society. 1962; **84**:4090-4095. DOI: 10.1021/ja00880a025

[14] Bronskill MJ, Taylor WB, Wolff RK, Hunt JW. Design and performance of a pulse radiolysis system capable of picosecond time resolution. The Review of Scientific Instruments. 1970;**41**: 333-340. DOI: 10.1063/1.1684511

[15] Tabata Y, Kobayashi H, Washio M, Tagawa S, Yoshida Y. Pulse radiolysis with picosecond time resolution. Radiation Physics and Chemistry. 1985;

26:473-479. DOI: 10.1016/0146-5724(85)90195-5

[16] Kobayashi H, Tabata Y. A 20 ps time resolved pulse radiolysis using two linacs. Nuclear Instruments and Methods: Section B. 1985;**10**:1004-1006. DOI: 10.1016/0168-583X(85)90158-2

[17] Yoshida Y, Tagawa S, Washio M, Kobayashi H, Tabata Y. Picosecond pulse radiolysis on geminate ion recombination and formation of solute excited state in liquid cyclohexane. Radiation Physics and Chemistry. 1989; **34**:493-496. DOI: 10.1016/1359-0197(89)90051-9

[18] Yoshida Y, Ueda T, Kobayashi T, Shibata H, Tagawa S. Studies of geminate ion recombination and formation of excited states in liquid n-dodecane by means of a new picosecond pulse radiolysis system. Nuclear Instruments and Methods A. 1993;**327**:41-43. DOI: 10.1016/0168-9002(93)91405-C

[19] Yoshida Y, Mizutani Y, Kozawa T, Saeki A, Seki S, Tagawa S, et al. Development of laser-synchronized picosecond pulse radiolysis system. Radiation Physics and Chemistry. 2001; **60**:313-318. DOI: 10.1016/S0969-806X(00)00368-6

[20] Kozawa T, Mizutani Y, Miki M, Yamamoto M, Suemine S, Yoshida Y, et al. Development of subpicosecond pulse radiolysis system. Nuclear Instruments and Methods: Section A. 2000;**440**: 251-253. DOI: 10.1016/S0168-9002(99)00997-3

[21] Wishart JF, Cook AR, Miller JR. The LEAF picosecond pulse radiolysis facility at Brookhaven National Laboratory. The Review of Scientific Instruments. 2004;**75**:4359-4365. DOI: 10.1063/1.1807004

[22] Muroya Y, Lin M, Han Z, Kumagai Y, Sakumi A, Ueda T, et al. Ultra-fast pulse radiolysis: A review of the recent system progress and its application to study on initial yields and solvation processes of solvated electrons in various kinds of alcohols. Radiation Physics and Chemistry. 2008;**77**:1176-1182. DOI: 10.1016/S0969-806X(00)00367-4

[23] Aoki Y, Nkajyo T, Tsunemi A, Yang J, Okada Y, Yorozu M, et al. Performance of compact pulse radiolysis system using a photocathode RF gun. Research on Chemical Intermediates. 2001;**27**:689-697. DOI: 10.1163/15685670152621951

[24] Marignier J-L, Waele V, Monard H, Gobert F, Larbre J-P, Demarque A, et al. Time-resolved spectroscopy at the picosecond laser-triggered electron accelerator ELYSE. Radiation Physics and Chemistry. 2006;**75**:1024-1033. DOI: 10.1016/j.radphyschem.2005.10.020

[25] Kawaguchi M, Ushida K, Kashiwagi S, Kuroda R, Kuibayashi T, Kobayashi M, et al. Development of compact picosecond pulse radiolysis system. Nuclear Instruments and Methods: Section B. 2005;**236**:425-431. DOI: 10.1016/j.nimb.2005.04.012

[26] Yang J, Kan K, Naruse N, Yoshida Y, Tanimura K, Urakawa J. Femtosecond pulse radiolysis and femtosecond electron diffraction. Nuclear Instruments and Methods in Physics Research A. 2011;**637**:S24-S29. DOI: 10.1016/j.nima.2010.02.014

[27] Yang J, Kondoh T, Kan K, Yoshida Y. Ultrafast pulse radiolysis. Nuclear Instruments and Methods in Physics Research A. 2010;**629**:6-10. DOI: 10.1016/j.nima.2010.11.109

[28] Yang J, Kondoh T, Kan K, Kozawa T, Yoshida Y, Tagawa S. Femtosecond single electron bunch generation by rotating longitudinal bunch phase space in magnetic field. Nuclear Instruments and Methods in Physics Research A. 2006;**556**:52-56. DOI: 10.1016/j.nima.2005.10.115

[29] Yang J, Kondoh T, Kozawa T, Yoshida Y, Tagawa S. Pulse radiolysis based on a femtosecond electron beam and a femtosecond laser light with double-pulse injection technique. Radiation Physics and Chemistry. 2006; **75**:1034-1040. DOI: 10.1016/j. radphyschem.2005.09.016

[30] Oulianov DA, Crowell RA, Gosztola DJ, Shkrob IA, Korovyanko OJ, Rey-de-Castro RC. Ultrafast pulse radiolysis using a terawatt laser Wakefield accelerator. Journal of Applied Physics. 2007;**101**:053102. DOI: 10.1063/1.2696204

[31] Yang J, Kondoh T, Yoshida A, Yoshida Y. Double-decker femtosecond electron beam accelerator for pulse radiolysis. The Review of Scientific Instruments. 2006;**77**:043302. DOI: 10.1063/1.2195090

[32] Kan K, Kondoh T, Yang J, Ogata A, Norizawa K, Yoshida Y. Development of double-decker pulse radiolysis. The Review of Scientific Instruments. 2012; **83**:073302. DOI: 10.1063/1.4731652

[33] Yang J, Kondoh T, Norizawa K, Yoshida Y, Tagawa S. Breaking time-resolution limits in pulse radiolysis. Radiation Physics and Chemistry. 2009; **78**:1164-1168. DOI: 10.1016/j. radphyschem.2009.07.017

[34] Wishart JF, Rao BSM. Recent Trends in Radiation Chemistry. New Jersey: World Scientific Publishing Co. Pte. Ltd.; 2010

[35] Yang J, Kondoh T, Yoshida Y, Tagawa S. Experimental studies of transverse and longitudinal beam dynamics in photoinjector. Japanese Journal of Applied Physics. 2005;**44**: 8702-8707. DOI: 10.1143/JJAP.44.8702

[36] Yang J, Sakai F, Yorozu M, Okada Y, Yanagida T, Endo A. Experimental studies of emittance growth and energy spread in a photocathode RF gun. Nuclear Instruments and Methods in Physics Research A. 2002;**491**:15-22. DOI: 10.1016/S0168-9002(02)01181-6

[37] Sakai F, Yang J, Yorozu M, Okada Y, Yanagida T, Endo A. Stable high-brightness electron beam system with a photocathode RF gun for short pulse X-ray generation by Thomson scattering. Japanese Journal of Applied Physics. 2002;**41**:1589-1594. DOI: 10.1143/JJAP.41.1589

[38] Yang J, Sakai F, Yanagida T, Yorozu M, Okada Y, Takasago K, et al. Low-emittance electron-beam generation with laser pulse shaping in photocathode radio-frequency gun. Journal of Applied Physics. 2002;**92**:1608-1612. DOI: 10.1063/ 1.1487457

[39] Nozawa I, Kan K, Yang J, Ogata A, Kondoh T, Gohdo M, et al. Measurement of < 20 fs bunch length using coherent transition radiation. Physical Review Accelerators and Beams. 2014;**17**:072803. DOI: 10.1103/ PhysRevSTAB. 17.072803

[40] Yang J, Yoshida Y. Ultrafast pulse radiolysis for observation of short-lived intermediate species in radiation chemistry. Radioisotopes. 2017;**66**(10): 395-406. DOI: 10.3769/radioisotopes. 66.395

[41] Elles CG, Jailaubekov AE, Crowell RA, Bradforth SE. Excitation-energy dependence of the mechanism for two-photon ionization of liquid H2O and D2O from 8.3 to 12.4eV. The Journal of Chemical Physics. 2006;**125**:044515. DOI: 10.1063/1.2217738

[42] Pimblott SM, Laverne JA, Bartels DM, Jonah CD. Reconciliation of transient absorption and chemically scavenged yields of the hydrated electron in radiolysis. The Journal of Physical Chemistry. 1996;**100**: 9412-9415. DOI: 10.1021/jp960816f

Chapter 4

Surface Plasmons and Optical Dynamics on Vanadium Dioxide

Hiroaki Matsui

Abstract

We report on plasmonic resonances on VO_2 nanodot arrays and associated optical dynamics. The plasmon excitations based on electric field interactions lead to red shifts of the plasmon resonances to lower photon energy with increasing nanodot size. The spectral linewidths of plasmon peaks gradually become narrow with increasing nanodot size. This is related to a reduction in plasmon damping with respect to the electronic band structure of VO_2. This specific band structure of VO_2 affects the optical dynamics of plasmon resonances at the sub-picosecond scale. The optical excitations of VO_2 comprise intraband and interband transitions. The existence of plasmon bands induces long-lived lifetimes on decay processes. Intraband transitions in the conduction band (C.B.) play an important role in producing long lifetimes, attributing to free carriers in the C.B. By contrast, interband transitions related to bound electrons contribute to plasmon damping. The dynamic optical responses are closely related to the electronic band structures of VO_2.

Keywords: VO_2, surface plasmon, infrared, dynamics, Mott insulator

1. Introduction

Recently, plasmonic materials based on oxide and compound semiconductors (e.g., ZnO, CuSe, and InN) have received much attention given that plasmonic responses can be tuned by external fields [1–5]. Investigation of these materials has led to the identification of a new family of plasmonic materials in the infrared (IR) range, which differ from noble metals (e.g., Ag and Au) that have fixed free electron densities. Oxide and compound semiconductors show ideal Drude terms in the IR range due to the absence of interband transitions in the band gap [6]. Plasmonic tuning can be effected by carrier injections due to control of the Fermi level in the electronic bands [7, 8]. The optical features arising from these emerging plasmonic semiconductors show promise for use in optical applications in the IR range.

Control of free carriers has been reported on oxide materials with strong electron-electron correlations. Of these, materials comprising vanadium dioxide (VO_2) show a sharp insulator-metal transition (IMT) based on Mott-related and Peierls-related processes [9], which can be controlled by external fields such as thermal, electrical, and optical inputs. In particular, dramatic change of the specific band structure of VO_2 with external fields provides more than a three orders of magnitude change in electrical conductance. The IMT character of VO_2 resulting from the spin-orbital interactions makes this material a candidate for use in optoelectronic applications [10–13]. The widespread interest in VO_2 has further focused on elucidating the complicated domain and grain structures in the vicinity of IMT. This determines the spatial and hysteretic nature of the phase transition of VO_2. Recent studies have reported optical phenomena from plasmonic structures of VO_2 with nanostructures such as nanodots and nanoparticles in an effort to understand the correlation between the optical response and phase transition [14–17]. These investigations have fueled further interest in nanostructures of VO_2 and are a very exciting and promising area of research with respect to active plasmonic materials.

The ultrafast dynamics of free carriers in metals have received much attention. Above all, photo-induced alternation of the plasmonic response is a powerful tool in the fabrication of optically controlled nanophotonic devices. The combination of ultrafast optical manipulations with large reflectance and transmittance is expected to yield promising results in applications concerning active plasmonics and optical switching [18, 19]. Thus far, plasmon modulations at the sub-picosecond scale have been reported on noble metals such as gold and silver, being the most commonly adopted plasmonic materials [20]. In general, effort aimed at exploiting ultrafast dynamics requires the production of intense concentrations of energy (hot electrons and hot holes) in the hosts. The dipolar excitations of all free carriers in plasmonic resonances have large extinction cross-sections, which can easily concentrate light into nanoscopic volumes. These phenomena have been utilized in furthering the development of certain areas of photo-chemistry such as photo-catalytic reactions and surface-enhanced Raman scattering (SERS).

Recently, investigations of the optical dynamics of oxide semiconductors have focused on Sn-doped In_2O_3 and Ga-doped ZnO [21, 22]. The noble metals have large amounts of free electrons in the host. Their plasmonic characters are hardly tunable. By contrast, oxide materials can be easily controlled by optical excitations such as intraband and interband transitions. For example, an interband transition is based on carrier excitation between the conduction band (C.B.) and valence band (V.B.), which can markedly change the plasma frequency that is dependent on carrier concentration, leading to an optical tuning of a plasmon response. However, an optical dynamics on an interband transition are known to show long plasmon lifetimes of several picoseconds. On the other hand, a carrier excitation (electron or hole) in the C.B. or V.B. is based on an intraband transition, resulting in significantly faster plasmon lifetimes at the sub-picosecond scale [23]. Therefore, oxide semiconductors are expected to show ultrafast optical modulations. Unfortunately, the plasmon dynamics of VO_2 have hitherto not been investigated. The electronic band structure of VO_2 differs largely from that of oxide semiconductors, which show different optical dynamics due to the strong electron-electron correlations of 3*d* bands.

In this chapter, we highlight the surface plasmons and associated optical dynamics of VO_2 with two-dimensional (2*D*) nanodot arrays. Periodic alignments of metallic nanodots can excite collective plasmon resonances due to electromagnetic coupling between nanodots that are in close proximity. This plasmon coupling in the nanodot arrays markedly influences the optical excitation and resonant energy. In the first section, we introduce surface plasmons of VO_2 nanodot arrays and their damping processes, which will contribute toward our understanding of ultrafast dynamics of plasmon resonances. In the second section, we report on the optical dynamics of VO_2 nanodot arrays by the plasmonic excitations based on intraband transitions of C.B. Ultrafast optical modulations are observed in VO_2 nanodot arrays at the sub-picosecond scale. This work provides new insight into plasmon-induced optical dynamics of VO_2.

2. Surface plasmons of VO_2 nanodot arrays

2.1 Fabrication of nanodot arrays

VO_2 nanodot arrays were fabricated by top-down processes as follows [24]. VO_2 films were epitaxially grown on Al_2O_3 (0001) substrates at a substrate temperature of 420°C using pulsed laser deposition. ArF laser pulses were focused on VO_2 targets in an oxygen atmosphere of 1.0 Pa. After growing VO_2 films, a UV nanoimprint resist with a thickness of 200 nm was spin-coated onto VO_2 film surfaces. The lithography process was performed using an UV lithography system to fabricate the nanodot array structure. The coated resists were pushed by quartz molds with UV transparency under a pressure of 3 MPa for 5 min. The residual resist on the VO_2 film surfaces after processing the UV lithography was removed by O_2 plasma irradiation. Additionally, reactive ion etching (RIE) using SF_6 plasma was used to process VO_2 films to obtain nanodot arrays. SF_6 RIE was conducted under an input of 60 W and a gas pressure of 1 Pa for 30 s. The resist layers that remained on VO_2 nanodot arrays were removed in toluene for completion of the VO_2 nanodots.

Figure 1 shows X-ray 2θ diffraction (XRD) patterns of VO_2 nanodot array samples with different sizes (D = 340, 600, and 816 nm). XRD patterns with the (020) diffraction plane are only shown. The patterns indicated a monoclinic structure corresponding to a low temperature phase of VO_2. The 490- and 716-nm nanodot array samples also exhibited similar XRD patterns.

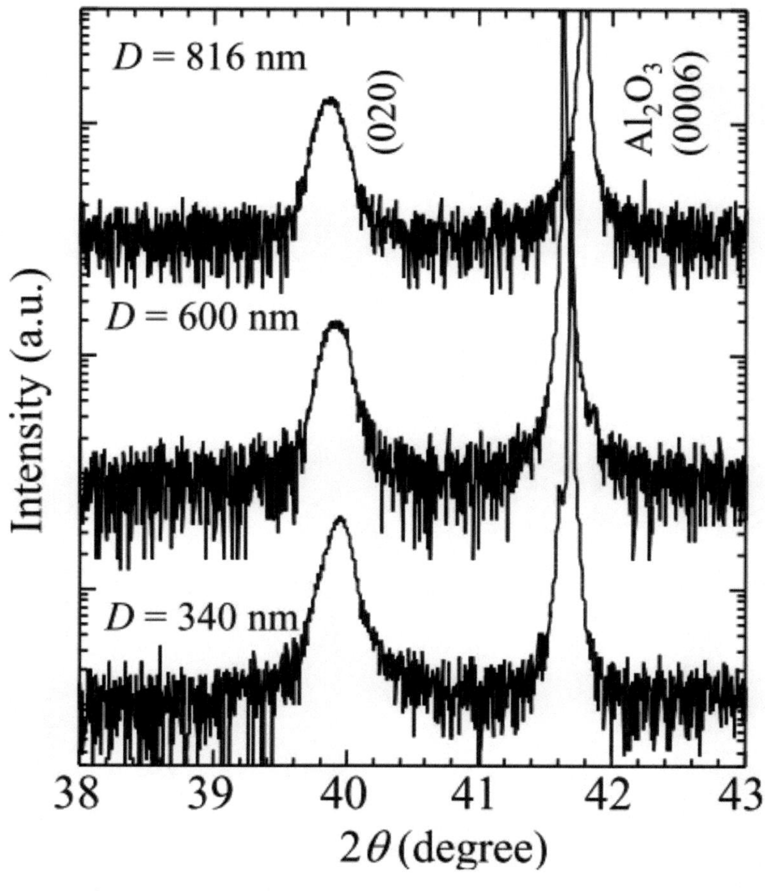

Figure 1.
2θ XRD patterns of VO_2 nanodot arrays with different sizes (D = 340, 600, and 816 nm) [24].

Temperature-dependent Raman measurements were investigated in an effort to identify structural changes in the VO_2 nanodot arrays (**Figure 2a**). The Raman peaks at 195, 224, and 614 cm^{-1} were mainly attributed to V-V (ω_1 and ω_2) and V-O (ω_o) vibrational modes, respectively [25]. Raman studies play an important role in clearing structural changes during the IMT, which is the index of our understanding regarding hysteresis phenomena of plasmon resonances. The peak at 195 cm^{-1} showed a slight change at 60°C. An additional peak at 632 cm^{-1} was detected above 60°C in the vicinity of the peak at 614 cm^{-1}. Insulating phases in VO_2 have the monoclinic M_1 and M_2 and triclinic T phases [26]. The M_2 phase is distinguished from the M_1 phase primarily by the red shift of the ω_o frequency [25]. **Figure 2a** shows that the M_1 phase at 60°C nucleated between the growing metallic rutile R phase and the remaining M_1 phase. The M_2 and R phases grew until the M_1 phase was consumed at this temperature and the R phase began to consume the M_2 phase. Furthermore, the transition to the R phase was complete at around 75°C. The hysteresis curve of the ω_0-related Raman peak shows that the hysteresis width was very broad (**Figure 2b**). Finally, we confirmed Raman peaks related to each nanodot using micro-Raman spectroscopy. All nanodots showed similar Raman spectra and associated temperature dependencies (**Figure 2c, d**). The XRD and Raman measurements revealed that the VO_2 nanodot arrays were well fabricated with high crystallinity.

2.2 Plasmon resonances and field distributions

Figure 3 shows the experimental and simulated extinction (A) spectra of VO_2 nanodot arrays with different sizes (D) from 340 to 816 nm, as characterized by $A = -\log_{10}(\text{transmittance})$. All samples were observed at 90°C in order to monitor plasmon resonances of the nanodot arrays. Extinction spectra were employed using two types of optical measurement systems. One comprised a Vis-NIR spectra from 2500 to 400 nm and the other is a FTIR spectra from 8000 to 4000 cm^{-1}. The

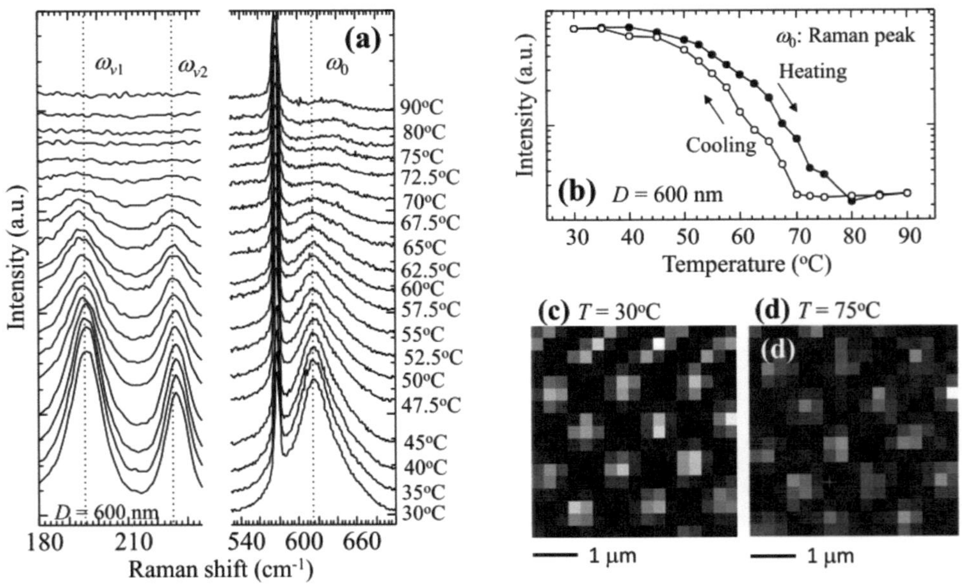

Figure 2.
(a) Temperature-dependent Raman spectra of the 600-nm nanodot array. (b) Hysteresis curve of the ω_o Raman active peak. Closed and open dots represent heating and cooling processes of the sample, respectively. 2D Raman images taken at (c) 30°C and (d) 75°C.

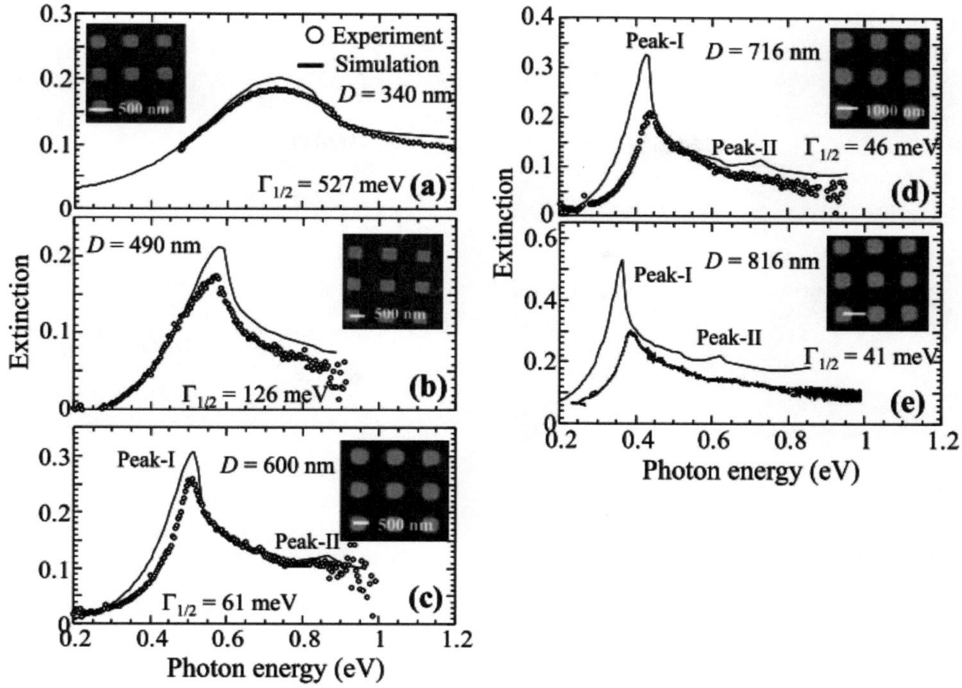

Figure 3.
Extinction (A) spectra and atomic force microscopy (AFM) images of nanodot arrays with different sizes [D = 340 (a), 490 (b), 600 (c), 716 (d), and 816 nm (e)]. Experimental and simulated spectra are represented by black lines and open circles, respectively [24].

nanodot size was checked using atomic force microscopy (AFM). The extinction spectrum of a D = 340 nm sample provided a broad spectral linewidth. However, the spectral linewidth gradually decreased with increasing nanodot size. The increase in nanodot size resulted in a red shift of the plasmon resonance to low photon energy, relating to long-range plasmon coupling based on the far-field since near-field interactions at a short range result in a blue shift of the resonance peak with increasing nanodot size [27]. Additionally, the experimental behavior was also observed with the simulated data using rigorous coupled-wave analysis (RCWA) (black lines in **Figure 3**). The magnitudes of the experimentally observed resonances were weaker than the simulated resonances, which may be a result of fabrication-related imperfections such as rounded edges and the corners of the nanodot arrays.

We could observe two kinds of resonance peaks (Peak-I and Peak-II) from the nanodot arrays with D = 600, 716 and 816 nm. Thereupon, we focused on the extinction of the nanodot array with D = 816 nm (**Figure 3e**), which showed Peak-I and Peak-II at 0.36 and 0.62 eV, respectively. The origin of the resonant peaks was clarified from three-dimensional field distributions based on the RCWA analysis. Simulations of the electric field (E-field) distributions were performed at 0.36 and 0.62 eV (**Figure 4**). The top-view E-fields at 0.36 eV (Peak-I) were enhanced at both interfaces between the nanodot and the substrate (substrate side), and between the nanodot and air (air side) (**Figure 4a, b**). The cross-section E-field revealed that the field intensity was more strongly concentrated on the substrate side than on the air side (**Figure 4c, d**), which showed a dipole plasmon resonance. On the other hand, the E-field at 0.62 eV (Peak-II) also emerged at both the substrate and air sides (**Figure 4e, f**), which was predominantly higher than that on the substrate side (**Figure 4g, h**). Therefore, the E-field at the substrate side adjacent to the substrate increased, and the field at the air side decreased. This behavior

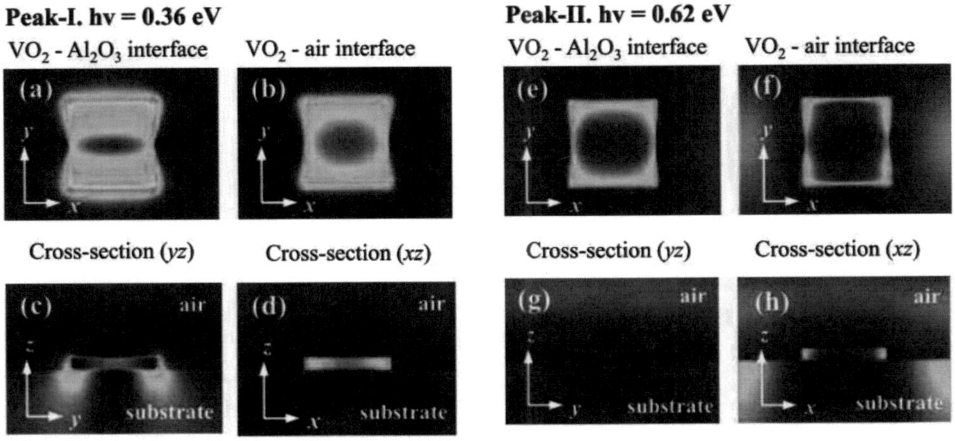

Figure 4.
Electric fields (E-field) of the nanodot array (D = 816 nm) calculated at resonances of 0.36 eV (upper images: Peak-I) and 0.62 eV (lower images: Peak-II). (a) and (e) show in-plane (x-y) field distributions at VO_2-Al_2O_3 interfaces at 0.36 eV and 0.62 eV, respectively. In-plane (x-y) field distributions of VO_2-air interfaces at 0.36 and 0.62 eV are exhibited in (b) and (f), respectively. (c) and (g) indicate cross-section (z-y) field distributions at 0.46 and 0.62 eV, respectively. Cross-section (z-x) field distributions at 0.36 and 0.62 eV are indicated in (d) and (h), respectively.

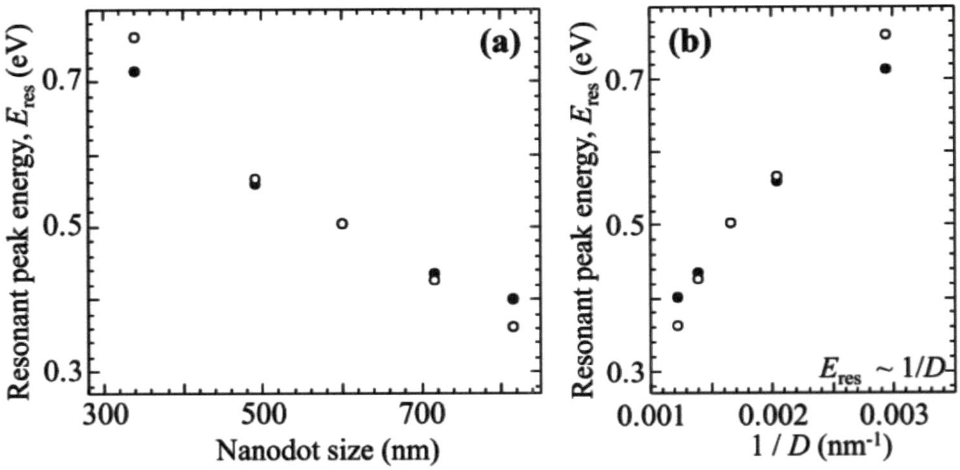

Figure 5.
(a) Energy positions of resonant peaks of Peak-I as a function of nanodot array. Experimental and simulated peaks are indicated by open and closed circles, respectively. (b) Scaling of the resonant peak energy (E_{res}) with inverse nanodot size (1/D) [24].

was consistent with the loss of symmetry following introduction of a substrate. The overall field distribution was weighted toward the lower edge in the vicinity of the substrate and toward the top edge away from the substrate. This phenomenon is attributed to substrate-induced plasmon hybridization for a metallic nanostructure on a dielectric substrate [28]. The dominant resonance of Peak-I, which is the focus of this study, mainly reflects the *E*-field between the nanodot and the substrate. This *E*-field provides the collective plasmon resonance in the nanodot arrays.

2.3 Damping processes

Figure 5a shows the relationship between nanodot size and resonant peak energy. The experimental and simulated data showed that a dominant peak energy

(E_{res}) associated with the dipolar resonance exhibited spectral red shifting with increasing nanodot size.

Besides, E_{res} was inversely proportional to $1/D$ (**Figure 5b**), which indicated that dynamic polarization effectively reduced the number of electrons oscillating coherently, and shifted the resonance energy into the low photon energy.

Figure 6a shows the spectral linewidth (Γ: FWHM) as a function of resonant peak energy (E_{res}). We extracted the spectral linewidth ($\Gamma = 2\Gamma_{1/2}$) as the red half-width at half-maximum $\Gamma_{1/2} = E_{res} - E_{1/2}$, where $E_{1/2}$ is the photon energy for which the extinction is half of the peak value [29]. The experimentally obtained Γ was systematically small with decreasing resonant energy, and was close to that of the simulated data. Plasmon quality ($Q = E_{res}/\Gamma$) enhanced with a reduction in resonant peak energy (**Figure 6b**). The slight difference in Γ between experimental and simulated data gradually increased with decreasing nanodot size. This may be related to surface damage of the nanodot owing to a side etching effect in the SF_6 plasma process.

The relationship between spectral linewidth and resonant energy is related to the dielectric functions of VO_2. **Figure 7a** shows complex dielectric functions of metallic VO_2 epitaxial films prior to nanofabrication. We discuss the plasmonic response of metallic VO_2 nanodots on the basis of bulk dielectric functions. The energy-dependent scattering rate (γ^*) and effective mass (m^*) are important factors in clarifying the damping mechanism, as estimated using the extended Drude model [30, 31].

$$\varepsilon(\omega) = \varepsilon_\infty - \frac{\omega_p^2}{\omega[\omega + i\gamma^*]} \tag{1}$$

where ε_∞ represents the high-energy dielectric constant derived from interband transitions. We used $\varepsilon_\infty = 9$. ω_p is the plasma frequency defined as follows [30, 31]: $\omega_p^2 = 4\pi Ne^2 / m^*(\omega)$,, where N_e is the total density of conduction electrons.

$$m(\omega)^* = \frac{4\pi e^2 N_e \gamma^*}{\varepsilon_2 \omega(\omega^2 + \gamma^{*2})} \tag{2}$$

$$\gamma(\omega)^* = \frac{\omega\varepsilon_2}{(\varepsilon_\infty - \varepsilon_1)} \tag{3}$$

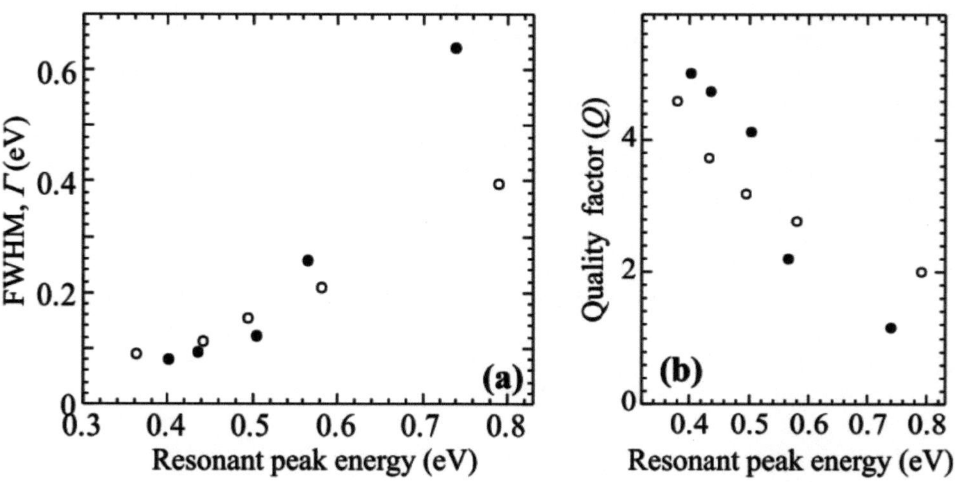

Figure 6.
(a) Plasmon linewidth (Γ) obtained from experimental (closed circles) and simulated (open circles) results as a function of resonant peak energy. Inset indicates an extinction spectrum of the nanodot array with D = *816 nm. (b) Plasmon quality* (Q = ħ/Γ) *as a function of* E_{res} *[24].*

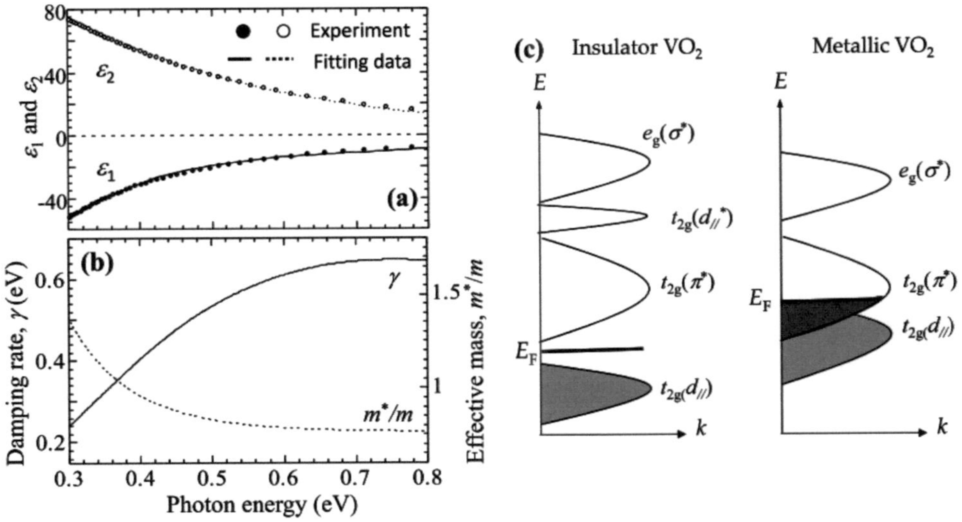

Figure 7.
(a) Real (ε_1: closed dots) and imaginary (ε_2: open dots) parts of the dielectric function in a VO_2 film prior to nanofabrications. Note that energy-dependent ε_1 and ε_2 in a VO_2 film were used since size-dependent plasmon damping is neglected. (b) Damping rate (γ^*: Straight line) and effective mass (m/m: dotted line) as a function of photon energy. (c) Schematic representation of electronic band structures of insulating and metal VO_2 [24].*

ε_1 and ε_2 indicate the real and imaginary parts of the dielectric function, respectively, when we define $\varepsilon\ (\omega) = \varepsilon_1(\omega) + i\varepsilon_2(\omega)$. Based on an N_e value of $5.2 \times 10^{22}\ cm^{-3}$ from a Hall measurement, the energy-dependent γ^* and m^*/m_0 (m_0 is the electronic band mass) in metal VO_2 were extracted (**Figure 7b**). In the simple Drude model, the scattering rate is considered independent of the photon energy. However, $\gamma\ (\omega)$ actually increased with increasing photon energy. The energy-dependent scattering rate is generally based on electron-electron (*e-e*) scattering and electron-phonon (*e-ph*) scattering. The behavior of $\rho(\omega)$ was proportional to ω^2, which indicated that the scattering rate of metal VO_2 was dominated by *e-e* scattering. Additionally, m^*/m_0 was enhanced slightly at a lower photon energy below 0.4 eV, which is characteristic of a Mott insulator [32]. Results showed that the energy-dependent Γ of plasmonic resonance was correlated with the energy-dependent γ in the extended Drude model. A plasmonic linewidth is mainly the sum of three separate processes involving intraband, interband, and radiative contributions (we ignore electron-surface damping, and chemical interface damping) [33]. The electronic band structure of VO_2 is explained as follows [34]. The crystal field splits degenerate 3*d*-orbitals into t_{2g} and e_g bands. The t_{2g} bands further split into $t_{2g}(d_{//})$ and $t_{2g}(\pi^*)$ bands. A $t_{2g}(d_{//})$ band splits into the upper $t_{2g}(d_{//}^*)$ and lower $t_{2g}(d_{//})$ bands in insulator VO_2. In metallic VO_2, an $t_{2g}(d_{//})$ band overlaps with an $t_{2g}(\pi^*)$ band at E_F, consisting only of 3*d* orbitals (**Figure 7c**). Additionally, radiation contribution to the plasmon damping was relatively small in this work since the plasmon linewidths gradually decreased with increasing nanodot size. As a consequence, intraband damping related to *e-e* scattering in the conduction band plays an important role in determining the plasmon linewidth, which is derived from the highly corrected electron system of VO_2.

3. Optical dynamics

Near-infrared laser pulses ranging from 1.9 to 2.5 μm were realized from a single supercontinuum pulse by intrapulse difference-frequency mixing with a spectral focusing scheme. For our pump and probe system, the probe wavelength was the same as that of the pump.

Details regarding the generation of femtosecond laser pulses (pulse width $\sim$ 150 fs) are described in [35].

We initially investigated differential transmittance ($\Delta T/T$) signals as a function of laser influence strength (P) (**Figure 8a**). $\Delta T/T$ can be defined as $\Delta T/T = (T_{on} - T_{off})/T_{off}$, with T_{on} and T_{off} being the probe signals reflected by the excited and unperturbed samples, respectively. An excitation wavelength of λ = 2.1 μm was used. The VO_2 nanodot array with D = 490 nm was used as a measurement sample. The sample temperature was maintained at 90°C to retain metallic state of VO_2. $\Delta T/T$ showed negative responses under different laser influence strengths (P = 100–560 mJ/cm^2), which indicated an increase in photo-induced absorbance. As shown in **Figure 8a**, the plasmon dynamics obey a well-known process [36]: carrier dephasing; electron-electron (*e-e*) scattering [process (1)], determined in the rise time, carrier cooling and lattice heating by electron-phonon (*e-ph*) scattering [process (2)], and followed by cooling of the lattice based on dissipation of phonons into the environment [process (3), photon-phonon (*ph-ph*) relaxation]. The optical dynamics of VO_2 nanodot arrays were mainly dominated by *e-e* and *e-ph* scattering processes in the plasmon relaxations. The rising (*e-e* scattering) and decay (*e-ph* scattering) time constants remained unchanged with different influence strengths, and were approximately 200 and 240 fs, respectively. The rising and decay processes could be described by a single exponential decay model. Besides, all $\Delta T/T$ signals were conducted below 1 ps, which differed from that employed for the noble metals [36]. The $\Delta T/T$ signals on silver and gold nanoparticles have commonly been extended in the range 10–20 ps [37]. Therefore, *e-e* and *e-ph* scattering on the VO_2 nanodot array was faster than that on the noble metals, contributing to optical modulations at the sub-picosecond scale. By contrast, *ph-ph* relaxation increased with increasing laser influence strength due to an increase in thermal relaxation.

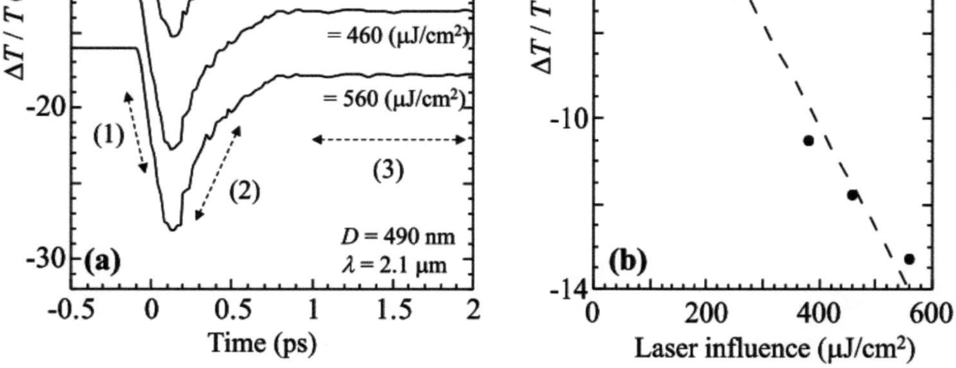

Figure 8.
(a) Differential transmittance (ΔT/T) signals as a function of lower influence strength. (1), (2), and (3) indicate e-e *scattering,* e-ph *scattering and* ph-ph *relaxation, respectively. (b) Dependence of* ΔT/T *on laser influence strength.*

The value of $\Delta T/T$ was inversely proportional to the laser influence strength (**Figure 8b**). No non-linear response was found within these laser influences.

On the other hand, no carrier dephasing was observed. The relaxation time of carrier dephasing is characterized by the time constant, T_2, which is related to the time constant for inelastic decay of the plasmon population, T_1, via $T_2^{-1} = T_1^{-1}/2 + T^{*-1}$, where T^* represents possible elastic phase-loss processes [38]. We can estimate T_2 using the following relation: $T_2 = 2\hbar/\Gamma$, where $\hbar$ and Γ (= $2\Gamma_{1/2}$) indicate the Plank constant and spectral linewidth, respectively. In the case of D = 490 nm, T_2 was estimated to be 5.1 fs, which was below the measurement limit.

Figure 9a shows differential transmittance signals as a function of laser wavelength for VO_2 nanodot arrays with D = 490 nm, revealing that the decay time of $\Delta T/T$ signals was dependent on laser wavelength. The decay time could be well fitted by a single exponential component in all laser wavelengths, as shown in **Figure 10a**. Photon energy-dependent plasmon lifetimes were extracted from decay times of $\Delta T/T$ signals. Lifetimes gradually increased with decreasing photon energy, and showed a maximum value (~0.257 ps) at 0.57 eV. The dependence of lifetime on photon energy was close to that on extinction. Similar results were obtained on the nanodot array with D = 600 nm (**Figures 9b** and **10b**). That is, the longest lifetimes were observed using laser wavelengths at the plasmon resonance peaks.

The extinction spectra of VO_2 nanodot arrays are characterized by the presence of plasmon bands in the IR range. The surface plasmon is a quantum of $h/(2\pi\omega_p)$ of plasma oscillation. The plasmon bands of VO_2 nanodot arrays were attributed to dipole oscillations of the free carriers (electrons) in the C.B. occupying the energy states immediately above the Fermi energy. The band structure of metallic VO_2 is more complex compared with oxide semiconductors (**Figure 7c**). The Fermi level is located in the $t_{2g}(\pi^*)$ level, which is partially overlapped with the $t_{2g}(d_{//})$ level. The optical excitations of metallic VO_2 comprised the three interband transitions E_1, E_2, and E_3 that were observed at around ~1.22, ~3.37, and ~5.90 eV, respectively [39]. In particular, the lowest optical transition E_1 corresponds to optical excitation from the filled $t_{2g}(d_{//})$ to empty $t_{2g}(\pi^*)$ bands. This band transition broadly exists in the range 0.5–2 eV, which is closely related to the plasmon dynamics.

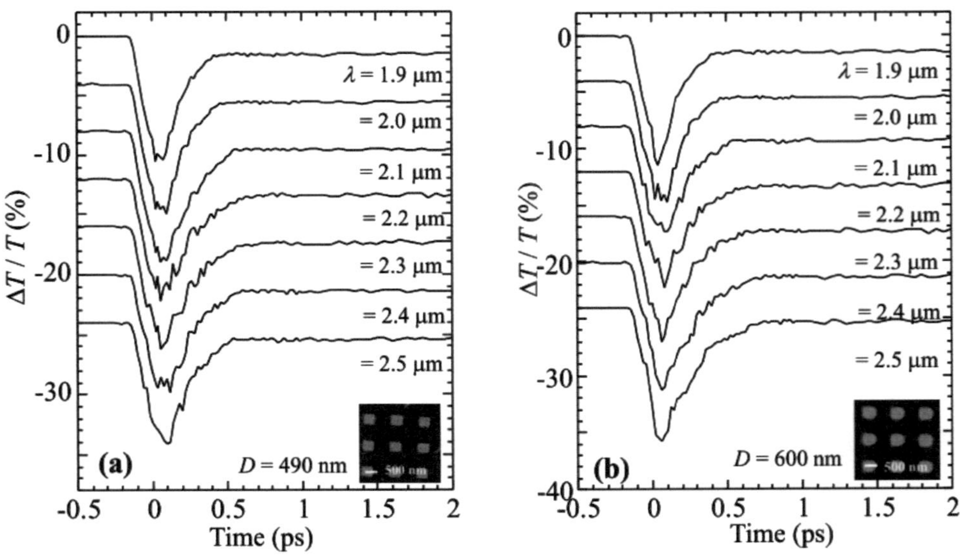

Figure 9.
Differential transmittance signals as a function of laser wavelength for VO_2 nanodot arrays with (a) D = *490 nm and (b)* D = *600 nm.*

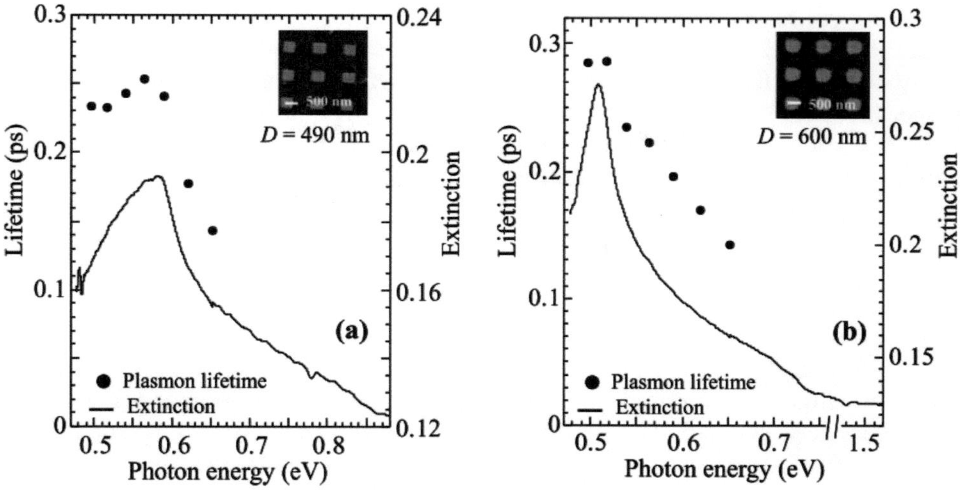

Figure 10.
Relationship between plasmon lifetime and extinction as a function of photon energy for VO_2 nanodot arrays with (a) D = 490 nm and (b) D = 600 nm.

In general, it is known that plasmon lifetimes differ largely between interband and intraband excitations. For an Au metal, the *s* band of the C.B. is overlapped with the *d* band, which provides two optical transitions as follows [40]. The interband excitations from *d* to *s* bands result in short-lived lifetimes due to the bounded carriers. By contrast, the intraband excitations in the C.B. consisting of the 6*sp* bands are composed of free carriers, leading to long-lived lifetimes as compared to the interband excitations (**Figure 11a**). On the other hand, for VO_2, the $t_{2g}(\pi^*)$ band is partially combined with the $t_{2g}(d_{//})$ band existing in VO_2 in addition to the

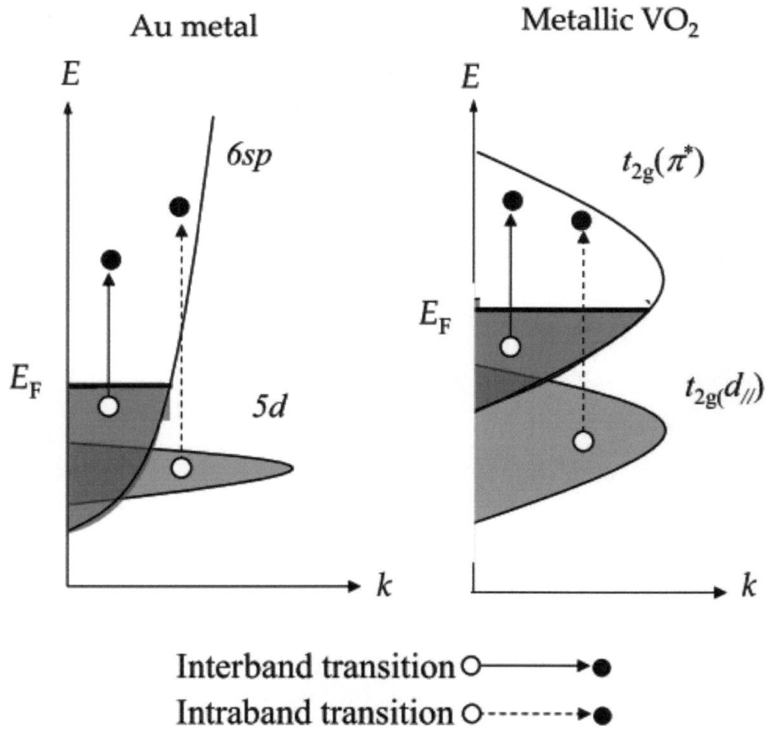

Figure 11.
The electronic band structures of Au metal (left) and metallic VO_2 (right).

intraband excitation within the $t_{2g}(\pi^*)$. That is, the long-lived lifetimes of plasmon peaks at 0.56 and 0.51 eV for nanodot arrays with D = 490 and 600 nm, respectively, were due to intraband excitations of free carriers in the C.B., which provided longer lifetimes than those of interband excitations. Optical dynamics of VO_2 were influenced by the intraband and interband transitions. The plasmon lifetimes were determined by the intraband excitations in the C.B., which were attributed to the band structure of VO_2.

4. Summary

We investigated infrared plasmonic responses of VO_2 nanodot arrays and their optical modulations. Comparison of the experimental plasmon resonances with electromagnetic simulations enabled us to perform spectral assignments and field distributions. In particular, plasmon coupling between metallic VO_2 nanodots contributed to the collective excitation mode. Plasmon damping of VO_2 was closely related to the specific band structure, which affected the optical dynamics. The plasmonic excitations excited by the fs pulse lasers showed ultrafast optical responses at the sub-picosecond scale, which were dependent on laser wavelength. The optical excitations of VO_2 comprised intraband and interband transitions. The long-lived lifetimes were observed at the resonant peaks in terms of free carrier excitations of VO_2. This result was attributed to the band structure, which affected the plasmon lifetimes.

Author details

Hiroaki Matsui[1,2]

1 Department of Bioengineering, The University of Tokyo, Bunkyo-ku, Tokyo, Japan

2 Department of Electrical Engineering and Information Systems, The University of Tokyo, Bunkyo-ku, Tokyo, Japan

*Address all correspondence to: hiroaki@ee.t.u-tokyo.ac.jp

References

[1] Matsui H, Furuta S, Tabata H. Mint: Role of electron carriers on local surface plasmon resonances in dope oxide semiconductor nanocrystals. Applied Physics Letters. 2014;**104**:211903. DOI: 10.1063/1.4880356

[2] Matsui H, Badalawa H, Hasebe T, Furuta S, Nomura W, Yatsui T, et al. Mint: Coupling of Er light emissions to plasmon modes on In_2O_3:Sn nanoparticle sheets in the near-infrared range. Applied Physics Letters. 2014; **105**:041903. DOI: 10.1063/1.4892004

[3] Runnerstrom EL, Bergerud A, Agrawal A, Johns RW, Dahkman CJ, Singh A, et al. Mint: Defect engineering in plasmonic metal oxide nanocrystals. Nano Letters. 2016;**16**:3390-3398. DOI: 10.1021/acs.nanolett6b01171

[4] Hsu SW, Ngo C, Tao AR. Mint: Tunable and directional plasmonic coupling within semiconductor nanodisk assemblies. Nano Letters. 2014;**14**: 2372-2380. DOI: 10.1021/nl404777h

[5] Palomaki PK, Miller EM, Neale NR. Mint: Control of plasmonic and interband transitions in colloidal indium nitride nanocrystals. Journal of the American Chemical Society. 2013;**135**: 14142-14150. DOI: 10.1021/ja404599g

[6] Naik CV, Shalaev VM, Boltasseva A. Mint: Alternative plasmonic materials: Beyond gold and silver. Advanced Materials. 2013;**25**:3264-3294. DOI: 10.1002/adma20125076

[7] Pattathil P, Giannuzzi R, Manca M. Mint: Self-powered NIR-selective dynamic windows based on broad tuning of the localized surface plasmon resonance in nanoporous ITO electrodes. Nano Energy. 2016;**30**:242-251. DOI: 10.1016/j.nanoen.2016.10.013

[8] Dahlman C, Agrawal A, Staller CM, Adair J, Milliron DJ. Mint: Anisotropic origins of localized surface plasmon resonance in n-type anatase TiO_2 nanocrystals. Chemistry of Materials. 2019;**31**:502-511. DOI: 10.1021/acs. chemmater.8b04519

[9] Kim DH, Kwok HS. Mint: Pulsed laser deposition of VO_2 thin films. Applied Physics Letters. 1994;**65**: 3188-3190. DOI: 10.1063/1.112476

[10] Basov DN, Averitt RD, Marel D, Deressel M, Haule K. Mint: Electrodynamics of correlated electron materials. Reviews of Modern Physics. 2011;**83**:471-541. DOI: 10.1103/ RevModPhys.83.471

[11] Kanki T, Hotta Y, Asakawa N, Kawai T, Tanaka H. Mint: Noise-driven signal transmission using nonlinearity of VO_2 thin films. Applied Physics Letters. 2010;**96**:242108. DOI: 10.1063/1.3455335

[12] Bae SH, Lee S, Lin L, Jo BH, Park C, Wang ZL. Mint: The memristive properties of a single VO_2 nanowire with switching controlled by self-heating. Advanced Materials. 2013;**25**:5098-5103. DOI: 10.1002/adma.201302511

[13] Zhang Z, Gao Y, Chen Z, Du J, Cao C, Kang L, et al. Mint: Thermochromic VO_2 thin films: Solution-based processing, improved optical properties, and lowered phase transformation temperature. Langmuir. 2010;**26**:10738-10744. DOI: 10.1021/la100515k

[14] Ke Y, Wen X, Zhao D, Che R, Xiong Q, Long Q. Mint: Controllable fabrication of two-dimensional patterned VO_2 nanoparticle, nanodome, and nanonet arrays with tunable temperature-dependent localized surface plasmon resonance. ACS Nano. 2017;**11**:7542-7551. DOI: 10.1021/acsnano.7b02232

[15] Zhu Z, Evans PG, Haglund RF Jr, Valentine JG. Mint: Dynamically

reconfigurable metadevice employing nanostructured phase-change materials. Nano Letters. 2017;**17**:4881-4885. DOI: 10.1021/acs.nanolett7b01767

[16] Appavoo K, Wang B, Seo M, Nag J, Prasankumar RP, Hilton DJ, et al. Mint: Ultrafast phase transition via catastrophic phonon collapse driven by plasmonic hot-electron injection. Nano Letters. 2014;**14**:1127-1133. DOI: 10.1021/nl4044828

[17] Zhu J, Huang A, Ma H, Ma Y, Tong K, Ji S, et al. Mint: Composite film of vanadium dioxide nanoparticles and ionic-nickel-chlorine complexes with excellent visible thermochromic performance. ACS Applied Materials & Interfaces. 2016;**43**:29742-29748. DOI: 10.1021/acsami.6b112020

[18] MacDonald KF, Sáimon ZL, Stockman MI, Zheludev NI. Mint: Ultrafast active plasmonics. Nature Photonics. 2009;**3**:55-58. DOI: 10.1038/NPHOTON.2008.249

[19] Harutyunyan H, Martinson ABF, Rosenmann D, Khorashad LK, Besteiro LV, Govorov AO, et al. Mint: Anomalous ultrafast dynamics of hot plasmonic electrons in nanostructures with hot spots. Nature Nanotechnology. 2015;**10**:770-775. DOI: 10.1038/NNANO2015.165

[20] Hodak JH, Martini I, Hartland GV. Mint: Spectroscopy and dynamics of nanometer-sized noble metal particles. Journal of Physical Chemistry B. 1998; **102**:6958-6967. DOI: 10.1021/jp9809787

[21] Guo P, Schaller RD, Ketterson JB, Chang RPH. Mint: Ultrafast switching of tunable infrared plasmons in indium tin oxide nanorod arrays with large absolute amplitude. Nature Photonics. 2016;**10**:267-273. DOI: 10.1038/NPHOTON.2016.14

[22] Calzolari A, Ruini A, Catellani A. Mint: Transparent conductive oxides as near-IR plasmonic materials: The case of Al-doped ZnO derivatives. ACS Photonics. 2014;**1**:703-709. DOI: 10.1021/ph500118y

[23] Diroll RT, Chang RPH, Schaller RD. Mint: Large transient optical modulation of epsilon-near-zero colloidal nanocrystals. ACS Nano. 2016;**10**: 10099-10105. DOI: 10.1021/acsnano.6b05116

[24] Matsui H, Ho YL, Kanki T, Tanaka H, Delaunay JJ, Tabata H. Mint: Mid-infrared plasmonic resonances in 2D VO_2 nanosquare arrays. Advanced Optical Materials. 2015;**3**:1759-1767. DOI: 10.1002/adom.201500322

[25] Atkin JM, Berweger S, Chavez EK, Raschke MB, Cao J, Fan W, et al. Mint: Strain and temperature dependence of the insulating phases of VO_2 near the metal-insulator transition. Physical Review B. 2012;**85**:02010(R). DOI: 10.1103/PhysRevB.85.020101

[26] Zhang S, Chou JY, Lauhon LJ. Mint: Direct correlation of structural domain formation with the metal insulator transition in a VO_2 nanobeam. Nano Letters. 2009;**9**:4527-4532. DOI: 10.1021/nl9028973

[27] Appavoo K, Lei DY, Sonnefraud Y, Wang B, Pantelides ST, Maier SA, et al. Mint: Role of defects in the phase transition of VO_2 nanoparticles probed by plasmon resonance spectroscopy. Nano Letters. 2012;**12**:780-786. DOI: 10.1021/nl203782y

[28] Zhang S, Bao K, Halas NJ, Xu H, Nordlander P. Mint: Substrate-induced Fano resonances of a plasmonic nanocube: A rote to increased-sensitivity localized surface plasmon resonance sensors revealed. Nano Letters. 2011;**11**:1657-1663. DOI: 10.1021/nl200135r

[29] Berciaud S, Cognet L, Tamarat P, Lounis B. Mint: Observation of intrinsic

size effects in the optical response of individual gold nanoparticles. Nano Letters. 2005;**5**:515-518. DOI: 10.1021/nl050062t

[30] Choi HS, Ahn JS, Jung JH, Noh TW, Kim DH. Mid-infrared properties of a VO_2 film near the metal-insulator transition. Physical Review B. 1996;**54**: 4621-4628. DOI: 10.1003/PhysRevB. 54.4621

[31] Makino H, Inoue IH, Rozenberg MJ, Hase I, Aiura Y, Onari S. Mint: Bandwidth control in a perovskite-type 3d—correlated metal $Ca_{1-x}Sr_xVO_3$. II. Optical spectroscopy. Physical Review B. 1998;**58**:4384-4393. DOI: 10.1003/PhysRevB.58.4384

[32] Okazaki K, Sugai S, Muraoka Y, Hiroi Z. Mint: Role of electron-electron and electron-phonon interaction effects in the optical conductivity of VO_2. Physical Review B. 2006;**73**:165116. DOI: 10.1103/PhysRevB.73.165116

[33] Hu M, Novo C, Funston A, Wang H, Staleva H, Zou S, et al. Mint: Dark-field microscopy studies of single metal nanoparticles: Understanding the factors that influence the linewidth of the localized surface plasmon resonance. Journal of Materials Chemistry. 2008;**18**: 1949-1960. DOI: 10.1039/B714759G

[34] Qazilbash MM, Schafgans AA, Burch KS, Yun SJ, Chae BG, Kim J, et al. Mint: Electrodynamics of the vanadium oxides VO_2 and V_2O_3. Physical Review B. 2008;**77**:115121. DOI: 10.1103/PhysRevB.77.115121

[35] Yamaguchi Y, Hida R, Suzuki T, Isa F, Yoshikiyo K, Fujii L, et al. Mint: Shaping and amplification of wavelength-tunable mid-infrared femtosecond pulses generated by intra-pulse difference-frequency mixing with spectral focusing. Journal of the Optical Society of America B. 2018;**35**:C1. DOI: 10.1364/JOSAB.35.0000C1

[36] Arbouet A, Voisin C, Christofilos D, Langot P, Del Fatti N, Vallée F, et al. Electron-phonon scattering in metal clusters. Physical Review Letters. 2003; **90**:177401. DOI: 10.1103/PhysRevLett. 90.177401

[37] Link S, Burda C, Wang ZL, El-Sayed MA. Electron dynamics in gold and gold-silver alloy nanoparticles: The influence of a nonequilibrium electron distribution and the size dependence of the electron-phonon relaxation. The Journal of Chemical Physics. 1999;**111**: 1255-1265

[38] Sönnichsen C, Franzl T, Wilk T, Plessen G, Feldmann J. Mint: Drastic reduction of plasmon damping in gold nanorods. Physical Review Letters. 2002;**88**:077402. DOI: 10.1103/PhysRevLett.88.077402

[39] Zhang P, Jiang K, Deng Q, You Q, Zhang J, Wu J, et al. Mint: Manipulations from oxygen partial pressure on the higher energy electronic transition and dielectric function of VO_2 films during a metal-insulator transition process. Journal of Materials Chemistry C. 2015;**3**:5033-5040. DOI: 10.1039/c5tc00002e

[40] Hartland GV. Mint: Optical studies of dynamics in noble metal nanostructures. Chemical Reviews. 2011;**111**:3858-3887. DOI: 10.1021/cr1002547

Chapter 5

Femtosecond Stimulated Raman Microscopy in C–H Region of Raman Spectra of Biomolecules and Its Extension to Silent and Fingerprint Regions

Rajeev Ranjan, Maria Antonietta Ferrara and Luigi Sirleto

Abstract

Stimulated Raman scattering (SRS) microscopy, based on vibrational spectroscopy, is able to perform *label-free* imaging with high sensitivity, high spatial and spectral resolution, 3D sectioning, and fast time of image acquisition, i.e., a few seconds. In this chapter, the implementation of a femtosecond SRS microscope, working in C–H or O–H region (>2800 cm^{-1}) of Raman spectra of biomolecules, is reported. Our microscope is realized by integration of a femtosecond (fs) SRS experimental setup with an inverted optical microscope equipped with fast mirror scanning unit. The microscope is provided with two femtosecond laser sources: a titanium-sapphire (Ti:Sa) laser oscillator and an optical parametric oscillator (OPO). In addition, the reliability of our system in C–H region is tested by localizations of lipid droplets inside adipocyte cells. Finally, the extension of microscope to silent region of <1800 cm^{-1} and fingerprint region of 1800–2800 cm^{-1} is also achieved by adding a second-harmonic generator to cascade the OPO and double its energy radiation. Definitely, our microscope is able to take on some recent challenges of SRS microscopy, including improvements of sensitivity and detection specificity.

Keywords: stimulated Raman scattering, optical imaging, nonlinear microscopy, label-free imaging, bio-imaging

1. Introduction

One of the most important subjects in life sciences is the noninvasive characterization of microscopic objects within a complex heterogeneous system by optical microscopy. Nowadays, fluorescence microscopy is the workhorse for cellular imaging. However, since most of small biomolecules, such as nucleosides, amino acids, fatty acids, choline, glucose, cholesterol, and small-molecule drugs, are intrinsically nonfluorescent, fluorescent tags are needed. Unluckily, these fluorescent probes, such as organic dyes, fluorescent proteins, or quantum dots, are all quite larger than small biomolecules; therefore they can severely falsify native biochemical or

IntechOpen

biophysical properties of the observed biomolecules inside cells. Definitely, small biomolecules have a fundamental functional importance, but, unfortunately, are basically undetectable by fluorescence microscopy. Therefore, a label-free imaging, with high sensitivity, high spatial resolution, and high chemical selectivity of unlabeled living cells, is highly desirable [1, 2].

Among label-free techniques, exploiting nonlinear optical effects, stimulated Raman scattering (SRS) microscopy is one of the most promising techniques [3–12]. SRS, discovered in 1960 [13–15], is a third-order nonlinear process in which two collinear laser pulses are used as pump and probe (Stokes) beams at frequencies of ω_p and ω_s (while $\omega_p > \omega_S$) and are focused into the sample. When the energy difference between pump and probe beams is equivalent to a vibrational energy (ω_v) of the sample, an exchange of energy between the two beams occurs and results a gain in the Stokes signal (stimulated Raman gain, SRG) and a loss in the pump signal (stimulated Raman loss, SRL) [16–19], as shown in **Figure 1**. Experimentally, both the modalities of SRG and SRL can be detected.

Since the invention of the picosecond (ps) narrowband SRS microscopy, a variety of molecular species have been imaged by label-free SRS targeting their intrinsic chemical bonds at the crowded cellular fingerprint region of 500–1700 cm^{-1} or the high-frequency C—H or O—H region of 2800–3200 cm^{-1}. However, SRS microscopy has some limitations, too. Indeed, by considering that several biomolecular species share similar chemical bonds, in some cases, it could be difficult to separate a target biomolecule from all the other related species inside cells. This leads to a problematic detection specificity. Additionally, the endogenous chemical bonds have typically small Raman scattering cross sections. This limits the detection sensitivity

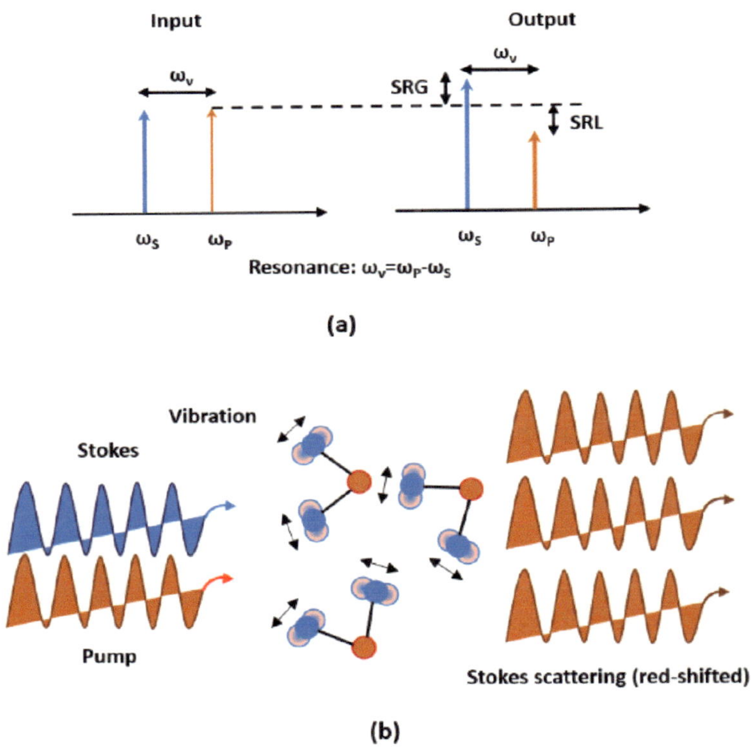

Figure 1.
SRS principle: (a) SRS modalities are described: SRG, stimulated Raman gain; SRL, stimulated Raman loss; and (b) stimulated Raman scattering occurs through the vibrationally excited molecules that interfere coherently.

of this approach to few millimolars, whereas a micromolar sensitivity is required for small biomolecules. Thus, there is a strong need to further increase specificity and sensitivity, as well as to probe dynamics [3–12].

Recently, to achieve detection specificity, general bio-orthogonal chemical imaging platform has emerged by coupling SRS microscopy with small Raman-active vibrational probes, such as alkynes and stable isotopes, which are both spectroscopically and biochemically orthogonal to the molecules inside cells, allowing a better sensitivity, specificity, and biocompatibility for imaging the small biomolecules. The improvement made by these vibrational probes is to shift Raman peaks in the cell-silent region where no other peaks from endogenous molecules are present [20–23].

Concerning SRS sensitivity, a 12 times larger stimulated Raman loss signal by femtosecond pulse excitation than picosecond pulse was demonstrated [24]. In literature, a number of interesting biological applications based on femtosecond SRS microscopy have been reported, extending to the silent region and encouraging the bio-orthogonal chemical imaging platform [25–30]. Definitely, femtosecond stimulated Raman microscopy and bio-orthogonal chemical appear as an unavoidable perspective for SRS improvements.

In multimodal imaging, single cells can be imaged in a single microscope by simultaneous image acquisition of various nonlinear imaging techniques: SRS, coherent anti-Stokes Raman scattering (CARS), second-harmonic generation (SHG), third-harmonic generation (THG), and sum frequency generation (SFG). The main advantages of multimodal imaging are that the same excitation source can be used for all imaging modalities, while generated signals by nonlinear interaction can be spectrally separated for simultaneous image collection in forward and/or backward detection, resolving structural diversity in biomaterials [31, 32].

In multimodal label-free microscopy, in order to take advantage of a single setup, several key laser parameters must be carefully considered. We note that SHG, THG, and multiphoton excitation fluorescence (MPEF) take advantage by short femtosecond pulses due to the nonlinear processes involved. For CARS and SRS, two synchronized pulses at the correct wavelength separation are required, so the laser system should have a wide spectral tunability in specific spectral regions. Therefore, in multimodal imaging to allow the most flexibility, the best choice are a femtosecond titanium-sapphire (Ti:Sa) oscillators and an optical parametric oscillator (OPO), where one of two laser beams can be used for single-beam techniques (SHG, THG, and MPEF), while their combination can be utilized for double-beam techniques (CARS or SRS) [31, 32].

However, although typical and widespread commercial laser sources, such as OPO pumped by Ti:Sa laser, are tailored for multimodal imaging, they present a limitation: Their minimum wavelength difference between pump (Ti:Sa laser) and signal (OPO) is 2500 cm^{-1}. Therefore, this laser combination allows only the CH-OH region exploration of Raman spectra, while the silent and fingerprint regions are out of the emission range [33–38]. As a consequence, this laser combination cannot accomplish the demand of bio-orthogonal platform based on femtosecond stimulated Raman microscopy, compromising its diffusion.

Recently, some new laser sources are available in the market. They allow to reach the silent region and fingerprint regions, but the pumping laser is a fiber one. It is well known that they have a relative intensity noise (RIN) greater than that of Ti:Sa laser oscillator, so a more complex detection system, for example, balanced detection, can be required. In this chapter in order to extend the potentiality of Ti:Sa laser and OPO combination and to avoid the drawback of fiber laser, an implementation of an SRS microscope is described, in which a third femtosecond source is added, i.e., a SHG cascading the OPO and doubling its energy radiation.

The chapter is organized as follows. In the next paragraph, the implementation and main realization issues of a femtosecond SRS microscope, working in C–H region, are illustrated. The microscope, equipped with two femtosecond laser sources, a Ti:Sa laser and an OPO output, covers the C–H region or O–H region (>2800 cm^{-1}) in stimulated Raman gain (SRG) modality. The successful implementation of our imaging system is demonstrated reporting label-free images of polystyrene beads with a diameter of 3 microns. In Subsection 2.1, SRS imaging is applied to investigate the distributions of lipid droplets (LDs) in 3T3-L1 cells at different stages of adipocyte differentiation process. In Section 3, the extension of previous SRS microscope to the silent and fingerprint regions, obtained by adding a third femtosecond laser source, i.e., a SHG cascading the OPO, is described. By combining the Ti:Sa laser and SHG output, the microscope covers not only the C–H region but also the fingerprint region (<1800 cm^{-1}) and the silent region (1800–2800 cm^{-1}) in stimulated Raman loss modality. Definitely, since in our proposed microscope, the femtosecond laser sources used are compatible with other nonlinear microscopy techniques, it could be integrated with multimodal microscopy.

2. SRS microscope in C–H region

It is worth noting that SRS microscopes are not commercially available, so in order to take advantage of their peculiarities, the only option is the homemade implementation. In order to set up an SRS microscope, two crucial issues have to be addressed:

- **Implementation of a high-frequency modulation transfer method:** SRS signal measures the pump-induced amplification of the Stokes pulse, detecting a tiny change in the transmitted probe light of the order of about 10^{-4}. Therefore, laser intensity noise and shot noise can reduce the SRS signal. In a high-frequency modulation transfer method, an electro-optic modulator (EOM) is used to modulate the pump and the modulation, is transferred to the probe beam, and can be detected by a photodiode (PD), after blocking the pump beam with a stack of optical filters. The PD output is connected by a low-pass filter to a lock-in amplifier (LIA), which demodulates the measured signal. SRS technique requires the extraction of a small alternative current (AC) signal at the sub-microvolt level from a noisy environment. A LIA is essentially a band-pass amplifier with variable central frequency and bandwidth, which allows to obtain a significant improvement in signal to noise ratio (S/N), using a phase sensitive detection. If the modulation frequency is faster than the typical laser noise (e.g., 1 MHz), high sensitivity detection is achieved at the moderate laser power required for biological and medical imaging [39–42].

- **Realization of two-dimensional (2D) images:** To obtain 2D images which can be achieved through beam scanning with 2D galvo mirrors and by a sequential collection of pixels, which are acquired and quantized in intensity by a data acquisition chain. The scanning mirrors (SMs) require three TTL signals that are made available by the microscope controller connected to the scan head unit: pixel clock, line sync, and frame sync. A 2D image is obtained by the synchronization of the forward detection unit with two SMs, which raster the focal spot across an x-y plane. The synchronization is achieved by managing (i) the PCI card (NI PCIe 6363) through an in-house LabVIEW program, (ii) the electrical signal detected by LIA, and (iii) the digital signals provided by the microscope scanning unit controller. All images with 512 × 512 pixels were analyzed with ImageJ software. The acquisition of a single image matrix takes 16 s. However, for a correct image acquisition, setting a number of parameters is required (e.g., pixel dimension and pixel dwell time). With the aim to preserve all information in an image, Nyquist's theorem must be followed during the sampling frequency, while to ensure the correct correspondence between spatial coordinates of pixels and measured SRS value in each pixel, the integration time of LIA have to be set equal or comparable to the pixel dwell time [43–46].

A fundamental procedure when carrying out SRS imaging is microscope alignment. The pulsed pump (Ti:Sa laser) and probe (synchronized OPO) lasers have to be carefully overlapped both in space and time at the image plane in order to generate a high SRS signal. Microscope alignments can be realized by three steps:

1. **Spatial overlap of two beams**. The initial step is the alignment of OPO and Ti:Sa lasers so that each beam arrives to the microscope. Then, we choose to consider the OPO as a reference beam. The Ti:Sa laser beam is spatially overlapped to OPO beam on a position-sensitive detector.

2. **Temporal overlap of two beams:** Considering that the pump and probe beams have slightly different beam paths inside the OPO housing, they show a time delay of about 5 ns at the OPO exit. Temporal overlap is usually obtained by a finely tuneable optical delay line. Usually, two techniques can be used to obtain the temporal overlap of two beams. The first can be obtained using a fast PD and oscilloscope, while the second is based on cross-optical correlator. In the case of the first technique, a rough overlap of two beams is obtained (uncertainty of 10 ps), whereas an accurate temporal overlap is obtained using a cross-correlator (resolution of 1 fs).

3. **Alignment of the two beams inside the microscope:** The sample is preliminary observed with white light to individuate the desired field of view (FOV). Then, laser beams, entering the microscope by a side port of microscope, are aligned to reach the PD mounted on the upstream. In order to optimize the FOV, the illumination, the focal position of microscope objectives, and checking if the two beams are spatially overlapped could be very useful to acquire transmission images (TI) for both Ti:Sa laser and OPO output.

The scanning microscope focuses the two laser beams into the sample by a microscope objective. After interaction with the sample, transmitted output pulses are collected by an upper objective. A PD and a LIA allow to measure the result of the SRS process occurring in the focal volume of the specimen after removing the pump pulses by a stack of optical filters. Our SRS microscope is shown in **Figure 2**. It is constructed by combining a femtosecond SRS spectroscopy setup with an inverted optical microscope (Eclipse TE-2000-E, Nikon) equipped with mirror scanning unit (C2, Nikon). The two pulsed laser sources are (i) a femtosecond Ti:Sa laser (Chameleon Ultra II), with a pulse duration of ~140 fs, a repetition rate of 80 MHz, and wavelengths of 680–1080 nm, and (ii) a femtosecond synchronized optical parametric oscillator (SOPO-Chameleon Compact OPO) pumped by a Ti:Sa laser. The pulse duration of SOPO output is ~200 fs. The repetition rate is 80 MHz, and the emission wavelength range is 1000–1600 nm. By combining the two laser systems, a minimum photon energy difference of 2500 cm^{-1} can be

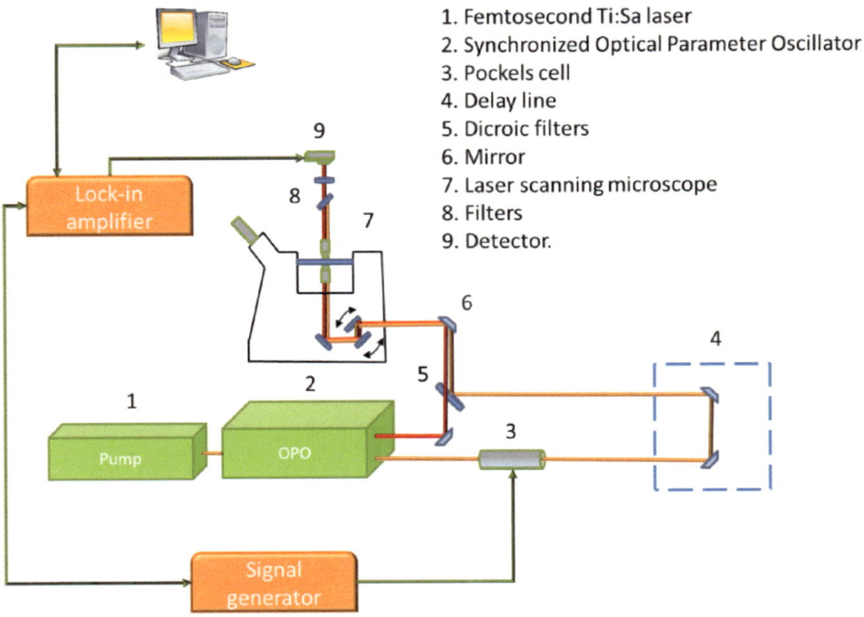

Figure 2.
Schematic experimental setup of the stimulated Raman scattering microscope.

obtained between the two beams. This means that only the high-frequency C-H region of Raman spectra (2800–3200 cm^{-1}) can be explored. Immediately following emission from the laser, the Ti:Sa laser pulses are modulated at a frequency of 4.5 MHz by an electro-optic modulator (EOM-CONOPTICS 350–160 KD*P) pulse selection system. Additionally, in order to obtain a temporal overlap between the two beams, a delay line (Newport MOD MILS200CC) was inserted between the Ti:Sa laser and the microscope. A dichroic mirror (Semrock FF875-Di01-125x36) spatially combines the collinear beams. The two beams were then focused into the specimen through a 60× multiphoton microscope objective (NA = 1.27). The output pulses are collected in transmission by a 40× high numerical aperture (NA = 1.25) multiphoton microscope objective. In order to remove the pump signal, a stack of optical filters was used, while the probe signal is measured by a photodetector (PD). The PD output is connected by a 50 Ω low-pass filter to a lock-in amplifier (LIA, SR844-200 MHz dual phase). The time constant in LIA was set to 100 μs with a slope of 18 dB/oct and a sensitivity of 10 μV [36, 37]. In order to collect information and to obtain a 2D image, the electrical signal coming out from the LIA is digitalized by PCI card (NI PCIe 6363), which manages and synchronizes the LIA and the scanning unit of microscope by an in-house developed LabVIEW program.

SRG images of a sample of polystyrene beads with a diameter of 3 μm were recorded. A strong SRG signal is observed from the polystyrene beads when the value of $\omega_L - \omega_S$ is matched to the C–H bond vibration at 3054 cm^{-1}. The obtained SRG and transmission images are shown in **Figure 3**. The pump signal was set at 810 nm, while the probe signal was set at 1073 nm. The focused power was less than 10 mW for both pump and probe beams. The transmission image was carried out with the single OPO laser at 1073 nm, and the transmitted beam intensity from the sample was measured by PD Thorlabs DET10N/M and directly acquired by the PCI card. The transmission image and the SRG image are given in **Figure 3(a)** and **(b)**, respectively.

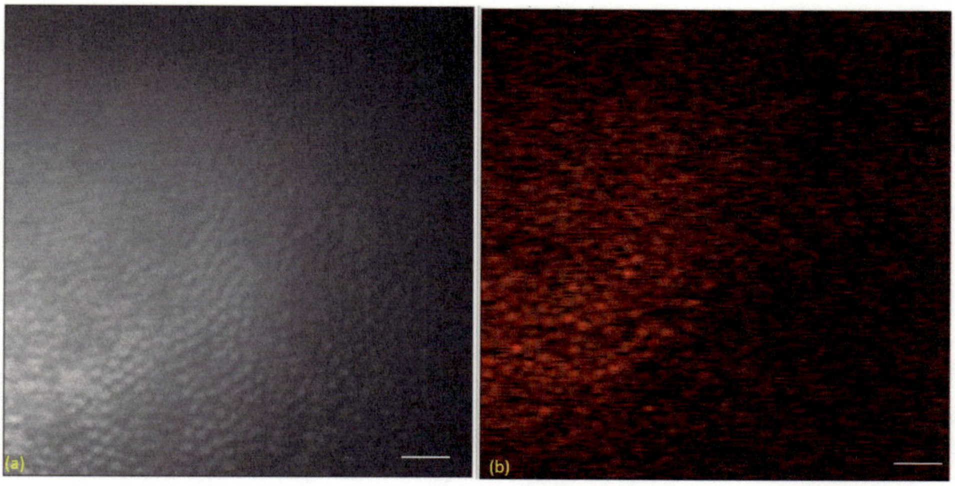

Figure 3.
(a) Transmission image and (b) SRG image of a polystyrene beads sample with a diameter of 3 μm. Scale bar corresponds to 25 μm.

2.1 Biological application in C–H region

Lipids are essential pillar for cells playing important roles for the maintenance of human health. Indeed, lipids act as signaling mediators, components of cellular membranes, and reservoirs for intracellular energy storage. Alterations in lipid metabolism are hallmarks of several human diseases, such as dyslipidemia, lipodystrophy, diabetes, obesity, atherosclerosis, heart diseases, and cancer.

The lipids are packed into specific intracellular organelles, termed lipid droplets, present in all cellular lines, but abundant in adipose and steroid-producing cells. LDs typically consist of neutral lipids in the form of triacylglycerols, cholesteryl esters, or retinyl esters encircled by a phospholipid monolayer with a surface made of numerous proteins that are involved in the regulation of lipid metabolism. Their sizes range from some tens of nanometer to tens of micrometer in diameter. LDs participate in a broad variety of physiological processes, such as lipid storage serving as fuels and/or such as biosynthetic precursors and signal transducers. Even though research is continuously growing in the relation between LDs and prominent diseases, a lot of fundamental questions are not yet resolved. In addition, new and unexpected connections with other cellular processes and pathologies are being found.

Traditionally LDs are imaged with fluorescence techniques of microscopy, using labelling fixed cells with neutral lipid specific dyes. However, lipids are smaller molecules than proteins and DNA, and their structures and functions are more susceptible to changes that might occur when adding fluorophores for imaging, introducing unwanted artifacts. Furthermore, their reluctant behavior to fluorescence dyes makes their identification and analysis difficult, especially for small LDs [47–60].

So far, most SRS applications have been aimed at label-free imaging of lipids in a variety of samples from artificial model systems to living cells and tissues. This prevalence is due to strong SRS signal for microscopy provided by LDs. Lipids are abundant in $C–H_2$ groups leading to a relatively isolated Raman peaks associated with C–H bond vibrational states at 2845 cm^{-1}, thus allowing to distinguish them from other chemical species. This peculiarity, together with the interest in the study of lipids and the possibility to image lipid structures, has encouraged SRS studies.

In the previous section, we have demonstrated the ability to acquire the images of the polystyrene beads in both transmission imaging and SRS modality of our system. In this section we are going to testify the biological application of SRS imaging technique. We have considered the distributions of LDs in 3T3-L1 cells at different stages of adipocyte differentiation: (i) 7 days and (ii) 15 days of growing time.

Adipocyte cells were grown on glass and fixed in PBS. To have a stable sample, the glass with the fixed cells is sandwiched between two coverslips. In order to obtain the image(s) of LDs inside adipocyte cells, the Raman shift of 2845 cm^{-1} associated with $C{-}H_2$ stretching mode has been investigated [40]. Hence, the Ti:Sa laser is set at 810 nm, and the wavelength of the OPO output is assigned to 1053 nm. After the power optimization of the laser beams (to avoid the thermal noises), the intensities of both lasers were set at 11 mW. Each acquired image is a single collection with 512 × 512 pixels. The acquisition time was 16 s. The LIA time constant was fixed to 100 μs with a slope of 16 dB/oct and a sensitivity of 10 μV.

Figure 4(a) shows the transmission image acquired image obtained with the OPO at wavelength of 1053 nm. The region of interest is first identified with this modality. **Figure 4(b)** shows the vibrational SRS contrast imaging. The bright regions indicate the individual lipid droplets. Small LDs are observable. The merge between the two images (a) and (b) is reported in **Figure 4(c)**. It highlights LD distribution inside the cells and its typical crown arrangement around the cell nucleus.

With the same procedure, we performed the analysis on a sample of adipocyte cells with 15 days of growing time. The results are shown in **Figure 5** for the transmission image obtained with OPO, SRS image, and the merge image.

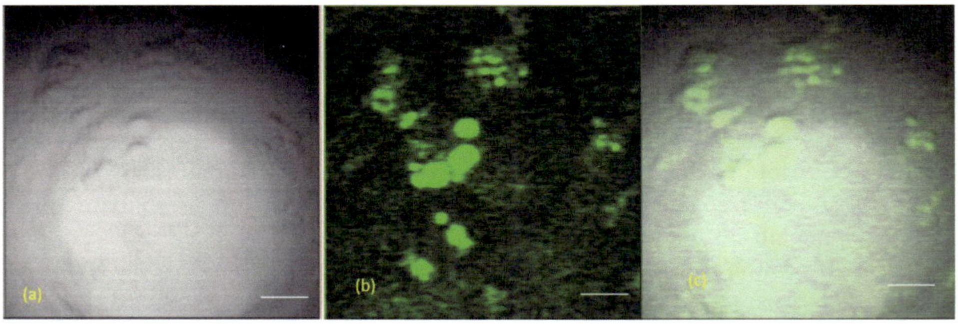

Figure 4.
Adipocyte cells at 7 days (D7): (a) transmission image by OPO, (b) SRS image of the sample, (c) merged image of transmission and SRS. Scale bar corresponds to 10 μm.

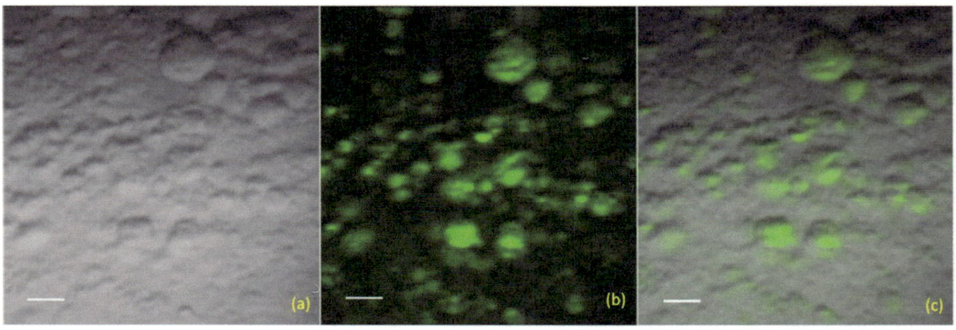

Figure 5.
Adipocyte cells at 15 days (D15) of growing time: (a) transmission image, (b) SRS image of the sample, and (c) merged image of absorption and SRS that indicates the LD distribution. Scale bar corresponds to 10 μm.

During our analysis in both the samples, we found that heterogeneity in LD size ranges from a few microns until 10 microns and beyond. Definitely, our findings indicate that SRS imaging is an alternative imaging tool when we cannot use fluorescence microscopy successfully.

3. Extension to the fingerprint region

The extension of the microscope to the silent region (<1800 cm^{-1}) and fingerprint region (1800–2800 cm^{-1}) is achieved by adding a SHG, cascading the OPO, and the SHG provides to double the OPO energy radiation. In our experimental setup, we used a flip mirror (FM) after the OPO exit to switch the Ti:Sa-OPO and Ti:Sa-SHG laser combinations. When the FM is inserted in the optical path, OPO radiation is coupled into SHG, so Ti:Sa-SHG laser combination becomes available, whereas when the FM is flipped, the OPO can go on undisturbed, and Ti:Sa-OPO laser combination becomes available (see **Figure 6**) [35].

Ti:Sa and OPO beams were spatially collinearly combined by a dichroic mirror (DM1) as described in the previous section. Ti:Sa-SHG beams were spatially collinearly combined by a dichroic mirror (DM2; Semrock FF735-Di0125x36), which is placed after DM1. Since DM2 is a high band-pass filter, whose transparence band starts at 735 nm, all wavelengths selected by the Ti:Sa-OPO laser combination are transmitted. Therefore, SRG acquisition in the Ti:Sa-OPO combination is not influenced by the presence of DM2.

Considering the Ti:Sa laser excitation with a wavelength range of 740–880 nm, the corresponding OPO emission range is 1000–1600 nm, while the SHG emission range is 500–800 nm. When the Ti:SA is used as pump at 830 nm, the largest range of Raman shift is obtained. Indeed, in this case all the region of Raman spectra can be covered (from 451 cm^{-1} to 3195 cm^{-1}) using the Ti:Sa-SHG laser combination.

The use of EOM in the implementation of SRS microscope leads to some drawbacks: increasing the cost of system, covering more space on the optical bench, and yielding unwanted electromagnetic interference. To avoid these disadvantages, save time for alignment, and increase flexibility of setup, the use of a single EOM with a fixed position on the bench is desirable. For these reasons, we implemented our SRS microscope with only one EOM. It is important to underline that the choice of laser source modulated by EOM has relevant consequences, because it settles the modalities SRL or SRG that can be utilized. We have chosen to use the EOM (350–160 KD*P CONOPTICS) for the intensity modulation of Ti:Sa laser pulses at a frequency *f* in the range of 1–30 MHz. For the Ti:Sa and OPO combination, only the SRG scheme can be thus implemented because the modulated laser beam energy is

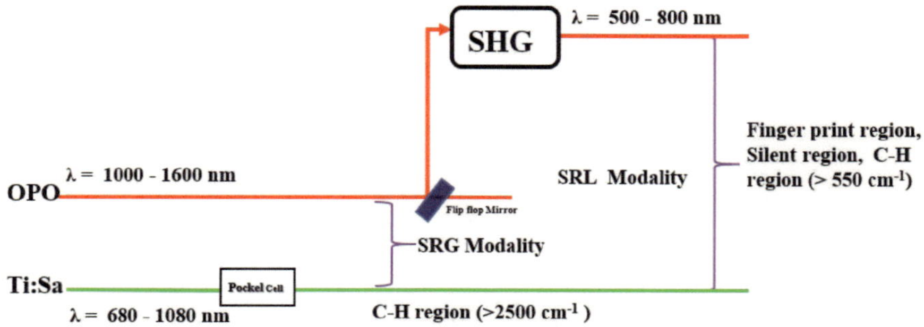

Figure 6.
Generation system with combination Ti:Sa-OPO (SRG modality) and Ti:Sa-SHG (SRL modality).

always greater than the other one. For the Ti:Sa-SHG combination, the SRL scheme can be implemented, since the modulated laser beam energy is always lower than the other one.

With the same scope to both reduce cost and save space, the use of a single delay line that allows coexistence between SRL and SRG optical circuitries is desirable. When the Ti:Sa-SHG laser configuration is used, the SHG only introduces an additional optical path with respect to Ti:Sa laser, with the delay time inside the SHG unchanged with respect to OPO. Therefore, in such configuration, temporal overlap between Ti:Sa and SHG laser pulses is achieved by a fixed shift movement of delay line with respect to the Ti:Sa-OPO overlap position. In our system, a single delay line with a fixed position on the bench is used, and the temporal overlap of each combination is achieved only by changing the direction of the delay line fixed shift movement, allowing a good flexibility (see **Figure 7**).

Again, in order to reduce cost and save space, a further important requirement is the use of a sole microscope for both optical circuits. In our setup, both laser combinations are then focalized and scanned on the sample. The output pulses are collected, as described in the previous section.

In particular, the collected radiations are incident onto a high band-pass optical filter (Dichroic filter FF875-Di01-25x36), whose transmission band starts at 875 nm, leading to two perpendicular optical paths. The OPO output is transmitted on the longitudinal path, while both Ti:Sa and SHG beams are reflected in the perpendicular path. Two photodiodes (PD1, PD2) positioned to the end of each path detect the unmodulated pulses (OPO beam for SRG modality and SHG beam for SRL modality), while the Ti:Sa modulated beam (pump pulses in SRG configuration and Stokes pulses in SRL configuration) is filtered. In order to enhance the acquisition efficiency, a Thorlabs detector (PD2; DET10N/M, InGaAs Detector) was used for SRG measurements, while a Thorlabs detector (PD1; DET10A/M, Si Detector) was used for SRL measurement. A stop line notch filter (Semrock NF03-808E-25) with band-pass of 788–828 nm is used to remove the Ti:Sa modulated beam. For SRL modality, a further filter (Semrock FF01–680/42–25) with band-pass 660–701 nm is added before photodiode to remove Ti:Sa residual pump beam. Finally, photodiode readout is demodulated by the same LIA introduced before to extract the modulation depth at the frequency *f*.

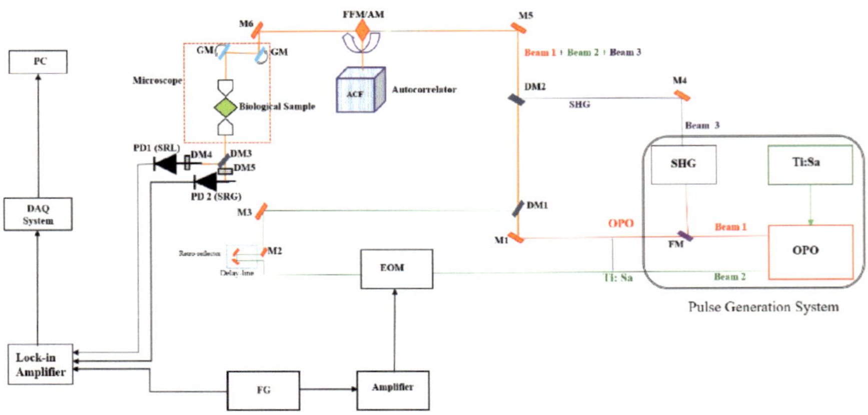

Figure 7.
Schematic layout of the f-SRS microscope. Ti:Sa, titanium-sapphire laser; OPO, optical parametric oscillator; SHG, second-harmonic generator; M1–M6, mirrors; FM, flip mirror; FFM/AM, flip flop mirror/ autocorrelator mirror; ACF, autocorrelator function DM1, DM2, DM3, dichroic mirror; DM4, DM5, notch filters; EOM, electro-optic modulator; FG, function generator; GM, galvo mirror; PD1, PD2, photodiode; DAQ, data acquisition system; PC, personal computer.

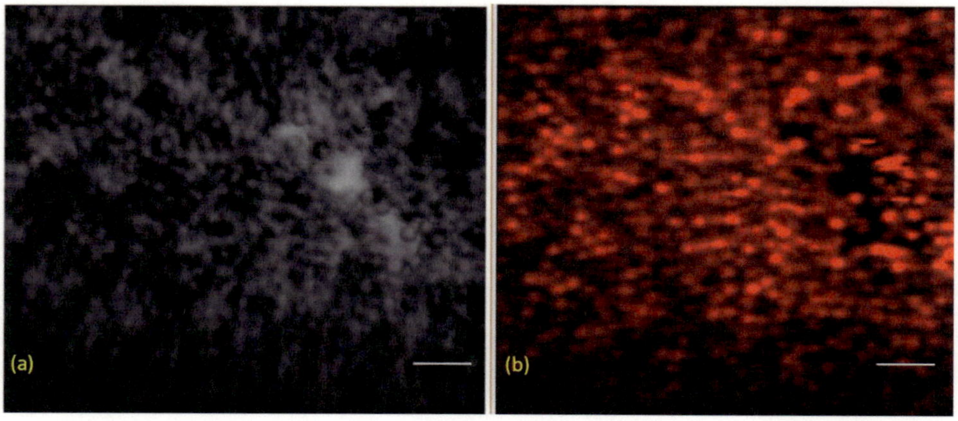

Figure 8.
(a) Transmission image and (b) SRL image of a polystyrene beads sample with a diameter of 3 μm. Scale bar corresponds to 25 μm.

To validate SRL imaging, the same PBS solution of polystyrene beads sample with diameter of 3 μm was used. The Raman shift was 3054 cm^{-1}, related to C–H bond. To record the image, the average focused power levels were 8 mW for pump beam at 650 nm generated by SHG unit and 10 mW for Stokes beam at 811 nm generated by Ti:Sa laser. The transmission image and the SRL image are shown in **Figure 8(a)** and **(b)**, respectively. In this case, the transmission image is obtained using a single beam at 650 nm. The transmitted beam intensity from the sample is measured by Thorlabs DET100A/M and directly acquired by the PCI card.

4. Conclusions

In this chapter, the design and the implementation of a nonlinear microscope based on stimulated Raman scattering is described. The presented microscope is equipped with three femtosecond laser sources: a titanium-sapphire laser, an optical parametric oscillator, and a second-harmonic generator (SHG) cascading the OPO. These three laser sources allow to cover all regions of Raman spectra by implementing two possible combinations. One consists in the use of Ti:Sa laser and OPO, which covers the C–H region or O–H region (>2800 cm^{-1}) in stimulated Raman gain (SRG) modality. Another with Ti:Sa laser and SHG covers not only the C–H region but also the fingerprint region (<1800 cm^{-1}) and the silent region (1800–2800 cm^{-1}) in stimulated Raman loss (SRL) modality. Therefore, our stimulated Raman scattering microscope allows the coexistence of SRG and SRL detection modes in a single microscopy setup and to acquire images of the same region in sequence without changing the optical components. Both SRL and SRG images of polystyrene beads are acquired by switching two combinations.

Finally, a biological case study in C–H region is reported, too. Our results confirmed that SRS imaging provides an advanced label-free approach to image and follow changes in lipid droplets potentially under physiological and pathological conditions.

Acknowledgements

The authors would like to thank Dr. G. Cozzi, a product specialist from Nikon Instruments, for his support in the SRS microscopy implementation and

Dr. M. Indolfi and Dr. V. Tufano (IMM-CNR) for their precious and constant technical assistance. The authors are also grateful to Angela Filograna and Carmen Valente of the Institute of Protein Biochemistry (CNR) for providing us with the adipocyte samples.

Conflict of interest

The authors declare that they have no competing financial interests.

Author details

Rajeev Ranjan[1,2]*, Maria Antonietta Ferrara[1] and Luigi Sirleto[1]

1 National Research Council (CNR), Institute of Applied Sciences and Intelligent Systems, Napoli, Italy

2 CHT@Erzelli, Nanoscopy, Istituto Italiano Di Tecnologia, Genova, Italy

*Address all correspondence to: rajeev.ranjan@iit.it

References

[1] Streets AM, Li A, Chen T, Huang Y. Imaging without fluorescence: Nonlinear optical microscopy for quantitative cellular imaging. Analytical Chemistry. 2014;**86**:8506-8513. DOI: 10.1021/ac5013706

[2] Min W, Freudiger CW, Lu S, Xie XS. Coherent nonlinear optical imaging: Beyond fluorescence microscopy. Annual Review of Physical Chemistry. 2011;**62**:501-530. DOI: 10.1146/annurev.physchem.012809.103512

[3] Popp J, Kiefer W. Raman scattering, fundamentals. In: Meyers RA, editor. Encyclopedia of Analytical Chemistry. Wiley; 2000. pp. 13104-13142

[4] Freudiger CW, Min W, Saar BG, Lu S, Holtom GR, He C, et al. Label-free biomedical imaging with high sensitivity by stimulated Raman scattering microscopy. Science. 2008;**322**(5909):1857-1861. DOI: 10.1126/science.1165758

[5] Saar BG et al. Video-rate molecular imaging in vivo with stimulated Raman scattering. Science. 2010;**330**(6009):1368-1370

[6] Zumbusch A, Langbein A, Borri P. Nonlinear vibrational microscopy applied to lipid biology. Progress in Lipid Research. 2013;**52**(4):615-632. DOI: 10.1016/j.plipres.2013.07.003

[7] Alfonso-García A, Mittal R, Lee ES, Potma EO. Biological imaging with coherent Raman scattering microscopy: A tutorial. Journal of Biomedical Optics. 2014;**19**(7):071407. DOI: 10.1117/1.JBO.19.7.071407

[8] Zhang D, Wang P, Slipchenko MN, Cheng JX. Fast vibrational imaging of single cells and tissues by stimulated Raman scattering microscopy. Accounts of Chemical Research. 2014;**47**(8):2282-2290. DOI: 10.1021/ar400331q

[9] Cheng JX, Xie XS. Vibrational spectroscopic imaging of living systems: An emerging platform for biology and medicine. Science. 2015;**350**(6264):aaa8870. DOI: 10.1126/science.aaa8870

[10] Camp CH Jr, Cicerone MT. Chemically sensitive bioimaging with coherent Raman scattering. Nature Photonics. 2015;**9**(5):295-305. DOI: 10.1038/nphoton.2015.60

[11] Fu D. Quantitative chemical imaging with stimulated Raman scattering microscopy. Current Opinion in Chemical Biology. 2017;**39**:24-31. DOI: 10.1016/j.cbpa.2017.05.002

[12] Lee HJ, Cheng J-X. Imaging chemistry inside living cells by stimulated Raman scattering microscopy. Methods. 2017;**128**:119-128. DOI: 10.1016/j.ymeth.2017.07.020

[13] Mc Clung FJ, Hellwarth RW. Giant optical pulsations from ruby. Journal of Applied Physics. 1962;**33**:828-829. DOI: 10.1364/AO.1.S1.000103

[14] Shen YR, Bloembergen N. Theory of stimulated Brillouin and Raman scattering. Physics Review. 1965;**137**:A1787-A1805. DOI: 10.1103/PhysRev.137.A1787

[15] Garmire E, Pandarese F, Townes CH. Coherently driven molecular vibrations and light modulation. Physical Review Letters. 1963;**11**:160-163. DOI: 10.1103/PhysRevLett.11.160

[16] Sirleto L, Vergara A, Ferrara MA. Advances in stimulated Raman scattering in nanostructures. Advances in Optics and Photonics. 2017;**9**(1):169-217. DOI: 10.1364/AOP.9.000169

[17] Ferrara MA, Sirleto L. Stimulated Raman scattering in micro- and nanophotonics. In: Lembrikov BI, editor. Nonlinear Optics - Novel Results in Theory and Applications. Rijeka: IntechOpen; 2018. DOI: 10.5772/intechopen.80814

[18] Sirleto L, Ferrara MA, Nikitin T, Novikov S, Khriachtchev L. Giant Raman gain in silicon nanocrystals. Nature Communications. 2012;**N.3**:1-6. DOI: 10.1038/ncomms2188, article number: 1220

[19] Sirleto L, Ferrara MA, Vergara A. Toward an ideal nanomaterial for on-chip Raman laser. Journal of Nonlinear Optical Physics and Materials. 2017;**26**. DOI: 10.1142/S0218863517500394

[20] Wei L, Hu F, Chen Z, Shen Y, Zhang L, Min W. Live-cell bioorthogonal chemical imaging: Stimulated Raman scattering microscopy of vibrational probes. Accounts of Chemical Research. 2016;**49**(8):1494-1502

[21] Hu F, Lamprecht MR, Wei L, Morrison B, Min W. Bioorthogonal chemical imaging of metabolic activities in live mammalian hippocampal tissues with stimulated Raman scattering. Scientific Reports. 2016;**6**(1):39660

[22] Wei L, Hu F, Shen Y, Chen Z, Yu Y, Lin CC, et al. Live-cell imaging of alkyne-tagged small biomolecules by stimulated Raman scattering. Nature Methods. 2014;**11**(4):410-412. DOI: 10.1038/nmeth.2878

[23] Wei L, Shen Y, Xu F, Hu F, Harrington JK, Targoff KL, et al. Imaging complex protein metabolism in live organisms by stimulated Raman scattering microscopy with isotope labeling. ACS Chemical Biology. 2015;**10**(3):901-908. DOI: 10.1021/cb500787b

[24] Zhang D, Slipchenko MN, Cheng JX. Highly sensitive vibrational imaging by femtosecond pulse stimulated Raman loss. Journal of Physical Chemistry Letters. 2011;**2**(11):1248-1253. DOI: https://doi.org/10.1021/jz200516n

[25] Zhang C, Li J, Lan L, Cheng JX. Quantification of lipid metabolism in living cells through the dynamics of lipid Droplets measured by stimulated Raman Scattering Imaging. Analytical Chemistry. 2017;**89**(8):4502-4507. DOI: 10.1021/acs.analchem.6b04699

[26] Dou W, Zhang D, Jung Y, Cheng J-X, Umulis DM. Label-free imaging of lipid-droplet intracellular motion in early Drosophila embryos using femtosecond-stimulated Raman loss microscopy. Biophysical Journal. 2012;**102**(7):1666-1675. DOI: 10.1016/j.bpj.2012.01.057

[27] Zhang C, Huang K-C, Rajwa B, Li J, Yang S, Lin H, et al. Stimulated Raman scattering flow cytometry for label-free single-particle analysis. Optica. 2017;**4**(1):103. DOI: 10.1364/OPTICA.4.000103

[28] Lee HJ, Zhang W, Zhang D, Yang Y, Liu B, Barker EL, et al. Assessing cholesterol storage in live cells and *C. elegans* by stimulated Raman scattering imaging of phenyl-diyne cholesterol. Scientific Reports. 2015;**5**(1):7930. DOI: 10.1038/srep07930

[29] Li J, Cheng J-X. Direct visualization of De novo Lipogenesis in single living cells. Scientific Reports. 2015;**4**(1):6807. DOI: 10.1038/srep06807

[30] Hu C-R, Zhang D, Slipchenko MN, Cheng J-X, Hu B. Label-free real-time imaging of myelination in the *Xenopus laevis* tadpole by in vivo stimulated Raman scattering microscopy. Journal of Biomedical Optics. 2014;**19**(8):086005. DOI: 10.1117/1.JBO.19.8.086005

[31] Zipfel WR, Williams RM, Webb WW. Nonlinear magic: Multiphoton microscopy in the biosciences. Nature Biotechnology. 2003;**21**(11):1369-1377. DOI: 10.1038/nbt899

[32] Hoover EE, Squier JA. Advances in multiphoton microscopy technology. Nature Photonics. 2013;7(2):93-101. DOI: 10.1038/nphoton.2012.361

[33] D'Arco A, Brancati N, Ferrara MA, Indolfi M, Frucci M, Sirleto L. Subcellular chemical and morphological analysis by stimulated Raman scattering microscopy and image analysis techniques. Biomedical Optics Express. 2016;**7**(5):1853-1864. DOI: 10.1364/BOE.7.001853

[34] D'Arco A, Ferrara MA, Indolfi M, Tufano V, Sirleto L. Label-free imaging of small lipid droplets by femtosecond-stimulated Raman scattering microscopy. Journal of Nonlinear Optical Physics and Materials. 2017;**26**(4):1750052. DOI: 10.1142/S0218863517500527.Jnopm

[35] Ranjan R, D'arco A, Ferrara MA, Indolfi M, Larobina M, Sirleto L. Integration of stimulated Raman gain and stimulated Raman losses detection modes in a single nonlinear microscope. Optics Express. 2018;**26**(20):26317-26326. DOI: 10.1364/OE.26.026317

[36] Ranjan R, Indolfi M, Ferrara, MA, Sirleto L. Implementation of a nonlinear microscope based on stimulated Raman scattering. Journal of Visualized Experiments. 2019;(149):e59614. DOI: 10.3791/59614

[37] Ferrara MA, Filograna A, Ranjan R, Corda D, Valente C, Sirleto L. Three-dimensional label-free imaging throughout adipocyte differentiation by stimulated Raman microscopy. PLoS One. 2019;**14**(5):e0216811. DOI: 10.1371/journal.pone.0216811

[38] Ranjan R, Ferrara MA, Filograna A, Valente C, Sirleto L. Femtosecond stimulated Raman microscopy: Home-built realization and a case study of biological imaging. Journal of Instrumentation. 2019;**14**:P09008. DOI: 10.1088/1748-0221/14/09/P09008

[39] Heritage JP, Allara DL. Surface picosecond Raman gain spectra of a molecular monolayer. Chemical Physics Letters. 1980;**74**(3):507-510. DOI: 10.1016/0009-2614(80)85262-6

[40] Levine BF, Shank CV, Heritage JP. Surface vibrational spectroscopy using stimulated Raman scattering. IEEE Journal of Quantum Electronics. 1979;**15**:1418-1432. DOI: 10.1109/JQE.1979.1069939

[41] Ozeki Y, Dake F, Kajiyama S, Fukui K, Itoh K. Analysis and experimental assessment of the sensitivity of stimulated Raman scattering microscopy. Optics Express. 2009;**17**(5):3651. DOI: 10.1364/OE.17.003651

[42] Nandakumar P, Kovalev A, Volkmer A. Vibrational imaging based on stimulated Raman scattering microscopy. New Journal of Physics. 2009;**11**:033026-033035. DOI: 10.1088/1367-2630/11/3/033026

[43] Qiang W, Merchant FA, Castleman K. Microscope Image Processing. San Diego, United States: Elsevier/Academic Press; 2008

[44] Serra J. Image Analysis and Mathematical Morphology. Orlando, FL, United States: Academic Press, Inc; 1993

[45] Gonzalez R, Woods R. Digital Image Processing. Upper Saddle River, NJ, USA: Prentice-Hall, Inc.; 2006

[46] Broeke J, Mateos PJ, Pascau J. Image Processing with ImageJ. Birmingham: Packt Publ; 2015

[47] Suzuki M, Shinohara Y, Ohsaki Y, Fujimoto T. Lipid droplets: Size matters. Journal of Electron Microscopy. 2011;**60**(1):S101-S116

[48] Yang L, Ding Y, Chen Y, Zhang S, Huo C, Wang Y, et al. The proteomics of lipid droplets: Structure, dynamics, and functions of the organelle conserved from bacteria to humans. Journal of Lipid Research. 2012;**53**(7):1245-1253. DOI: 10.1194/jlr R024117

[49] Thiam AR, Farese RV, Walther TC. The biophysics and cell biology of lipid droplets. Nature Reviews. Molecular Cell Biology. 2013;**14**(12):775-786. DOI: 10.1038/nrm3699

[50] Tirinato L, Pagliari F, Limongi T, Marini M, Falqui A, Seco J, et al. An overview of lipid droplets in cancer and cancer stem cells. Stem Cells International. 2017:1656053. DOI: 10.1155/2017/1656053

[51] Kinkel DA, Fernyhough ME, Helterline DL, Vierck JL, Oberg KS, Vance TJ, et al. Oil red-O stains nonadipogenic cells: a precautionary note. Cytotechnology. 2004;**46**:49-56. DOI: 10.1007/s10616-004-3903-4

[52] Ohsaki Y, Shinohara Y, Suzuki M, Fujimoto T. A pitfall in using BODIPY dyes to label lipid droplets for fluorescence microscopy. Histochemistry and Cell Biology. 2010;**133**:477-480. DOI: 10.1007/s00418010-0678-x

[53] Rizzatti V et al. Lipid droplets characterization in adipocyte differentiated 3T3-L1 cells: Size and optical density distribution. European Journal of Histochemistry. 2013;**57**(3):159-162. DOI: 10.4081/ejh.2013.e24

[54] Alfonso Garcia A et al. D38-cholesterol as a Raman active probe for imaging intracellular cholesterol storage. Journal of Biomedical Optics. 2016;**21**(6), article number: 061003-1:8. DOI: 10.1117/1.JBO.21.6.061003

[55] Mukherjee S, Zha X, Tabas I, Maxfield FR. Cholesterol distribution in living cells: Fluorescence imaging using dehydroergosterol as a fluorescent cholesterol analog. Biophysical Journal. 1998;**75**(4):1915-1925. DOI: 10.1016/S0006-3495(98)77632-5

[56] Fukumoto S, Fujimoto T. Deformation of lipid droplets in fixed samples. Histochemistry and Cell Biology. 2002;**118**(5):423-428. DOI: 10.1007/s00418-002-0462-7

[57] Kinkel A et al. Oil red-O stains non-adipogenic cells: A precautionary note. Cytotechnology. 2004;**46**(1):49-56. DOI: 10.1007/s10616-004-3903-4

[58] Farese RV, Walther TC. Lipid droplets finally get a little R-E-S-P-E-C-T. Cell. 2009;**139**:855-886. DOI: 10.1016/j.cell.2009.11.005

[59] Straub BK, Herpel E, Singer S, Zimbelmann R, Breuhahn K, Macher-Goeppinger S, et al. Lipid droplet-associated PAT-proteins show frequent and differential expression in neoplastic steatogenesis. Modern Pathology. 2010;**23**:480-492. DOI: 10.1038/modpathol.2009.191

[60] Pol A, Gross SP, Parton RG. Review: biogenesis of the multifunctional lipid droplet: Lipids, proteins, and sites. The Journal of Cell Biology. 2014;**204**(5):635-646. DOI: 10.1083/jcb.201311051

Chapter 6

Diffraction by a Rectangular Hole in a Thick Conducting Screen

Hirohide Serizawa

Abstract

The phenomenon of diffraction by a rectangular hole in a thick conducting screen is investigated for various structural parameters (aperture sizes, aspect ratios, and screen thicknesses) and some incident angles by making use of the exact solution based on the Kobayashi potential (KP) when an electromagnetic (EM) plane wave with any polarization is impinged on the aperture. Since the KP method yields an eigenfunction expansion of the present geometry, the solution satisfies the proper edge condition as well as all the boundary conditions, and therefore we can obtain highly accurate and fast-convergent results. Many numerical results, which are useful for scientists and engineers, are provided for various physical quantities, such as the far-field diffracted pattern, transmission coefficients (normalized transmitted power), and aperture electric field distributions, and by using these numerical results, we examine the convergent property of the KP solution and discuss the effect of the hole size and shape, screen thickness, and incident polarization on the transmission property of the rectangular hole.

Keywords: electromagnetic wave, exact solution, rectangular hole, thick screen, diffraction, transmission, aperture field

1. Introduction

A rectangular hole in a conducting screen with nonzero thickness is a fundamental structure and its diffraction property of an electromagnetic (EM) wave is of great importance in many fields of engineering and physics. For example, the knowledge of EM wave transmission through and radiation from the rectangular aperture with a finite thickness helps to design an aperture-type antenna like a slot antenna and to solve the problems of electromagnetic compatibility (EMC)/electromagnetic interference (EMI). The first theoretical work on this problem (three-dimensional problem) was made by Jin and Volakis by combining the finite element and the boundary integral methods [1]. From then, this problem has been solved by a variety of methods, such as the Fourier transform and mode-matching technique [2], a technique based on a perturbation method and modified Green's functions [3] and so on. In the field of physics, the extraordinary optical transmission (EOT) phenomena for subwavelength apertures have attracted considerable attention, and the study of the optical properties of holes in metallic films has become extremely active [4]. Garcia-Vidal et al. have applied the method based on the mode-expansion and mode-matching technique with some approximations to examine the transmission property of a single rectangular hole in a screen made of a perfect electric conductor (PEC) [5] and of a real metal (with surface impedance) [6]. A similar modal method (but without any approximation in the formulation) was used by Brok and Urbach to calculate the transmission through holes [7]. Many methods have thus been used to solve the EM wave diffraction by a rectangular hole in a thick screen, but all the methods mentioned above do not consider the field singularity at aperture edges in the formulation (e.g., EM fields near the edge of a deep hole are considered to behave like those around the right-angled wedge). It is well known that the incorporation of the proper edge condition [8] into the field is effective in obtaining a highly accurate and faster convergent solution, and the method that can take into account the edge property is required to obtain the exact solution. The method of the Kobayashi potential (KP) [9] is a rigorous technique for solving mixed boundary value problems and has been successfully applied to many three-dimensional wave scattering and radiation problems to obtain exact solutions [10–14]. The KP method uses the discontinuous properties of the Weber-Schafheitlin (WS) integrals to satisfy a part of the boundary conditions, and, at this step, the required edge condition can be incorporated into the solution. Serizawa and Hongo applied the KP method to the problem of diffraction of an EM plane wave by a rectangular hole in a PEC screen with a finite thickness and derived the exact solution of the diffracted wave that satisfy the proper edge condition as well as all the boundary conditions [15]. By using the derived KP solution, the physical quantities such as the transmission coefficient, far diffracted fields, and power flow around the hole have been calculated for small and large apertures [16–18], but the numerical results for apertures greater than the half-wavelength in [16] contain errors because of the mistake in the calculation code of double infinite series.

In this chapter, using the KP solution, we provide many numerical results of various physical quantities useful for scientists and engineers, and we investigate the convergent property of the KP solution and the transmission property of the rectangular hole in the thick conducting screen for various structural parameters (aperture sizes, aspect ratios, and screen thicknesses) and some incident angles (all the numerical results in this chapter are newly calculated by using the improved calculation code that can give more accurate results than the previous code).

2. Formulation

Consider an EM plane wave diffracted by a $2a \times 2b$ size rectangular hole in an infinitely large conducting screen with a finite thickness d as shown in **Figure 1**. The screen is made of a PEC, and the center of the upper aperture is chosen as the origin O of the Cartesian coordinates (x, y, z). The space is filled with an isotropic homogeneous medium with parameters (ϵ, μ). Through this chapter, a harmonic time dependence $\exp(j\omega t)$ is assumed and omitted from the equations. The expression of the incident wave (the angles of incidence are θ_0 and ϕ_0) is given by

$$\mathbf{E}^{inc} = (E_1 \mathbf{i}_\phi + E_2 \mathbf{i}_\theta) \exp\left[jk\Phi^{inc}(\mathbf{r})\right] \tag{1}$$

$$\mathbf{H}^{inc} = Y(E_1 \mathbf{i}_\theta - E_2 \mathbf{i}_\phi) \exp\left[jk\Phi^{inc}(\mathbf{r})\right] \tag{2}$$

where $k = \omega\sqrt{\epsilon\mu}$, $Y = \sqrt{\epsilon/\mu}$, and

$$\mathbf{i}_\theta = \cos\theta_0 \cos\phi_0 \mathbf{i}_x + \cos\theta_0 \sin\phi_0 \mathbf{i}_y - \sin\theta_0 \mathbf{i}_z \tag{3}$$

$$\mathbf{i}_\phi = -\sin\phi_0 \mathbf{i}_x + \cos\phi_0 \mathbf{i}_y \tag{4}$$

$$\Phi^{inc}(\mathbf{r}) = x \sin\theta_0 \cos\phi_0 + y \sin\theta_0 \sin\phi_0 + z \cos\theta_0. \tag{5}$$

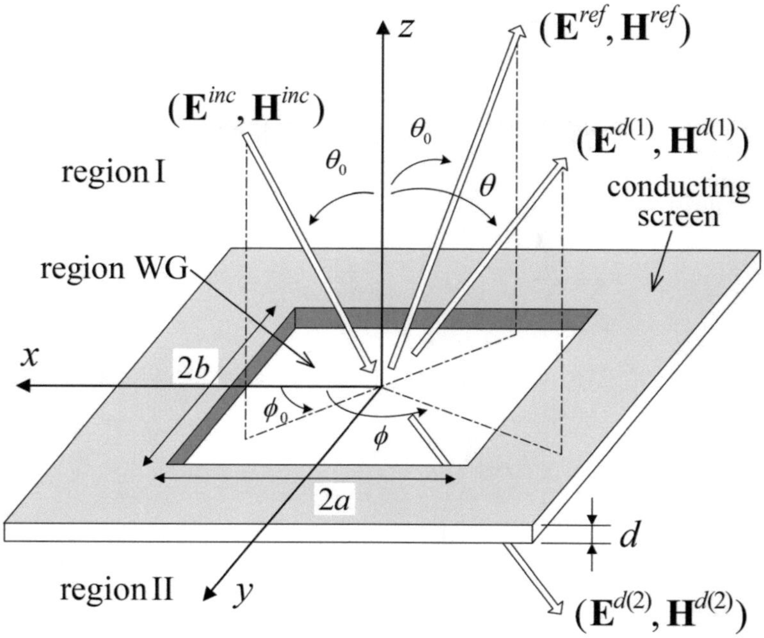

Figure 1.
Plane wave diffraction by a rectangular hole in a perfectly conducting infinite screen with a finite thickness d. The hole size is 2a × 2b, and the center of the upper aperture is chosen as the origin of the Cartesian coordinates.

In **Figure 1**, $\mathbf{E}^{ref}$ is the reflected wave when the plane $z = 0$ is occupied by an infinite conducting screen.

2.1 Field expression in each region

To solve the problem, we split the space into three regions I, II, and WG, denoting the upper and lower half-spaces, and waveguide region, respectively. Needless to say, $\mathbf{E}^{d(2)}$ is the transmitted wave.

The fields in region WG are represented by a linear combination of the TE- and TM-modes, and the axial components of the electric and magnetic vector potentials are given by

$$\begin{pmatrix} F_z^+ \\ F_z^- \end{pmatrix} = a\epsilon \sum_{\substack{m=0 \\ (m,n)\neq(0,0)}}^{\infty} \sum_{n=0}^{\infty} \begin{pmatrix} A_{mn}^{(E)} \exp\left(-jh_{mn}z_a\right) \\ B_{mn}^{(E)} \exp\left(jh_{mn}z_a\right) \end{pmatrix} \cos\frac{m\pi}{2}(\xi+1)\cos\frac{n\pi}{2}(\eta+1) \quad (6)$$

$$\begin{pmatrix} A_z^+ \\ A_z^- \end{pmatrix} = \frac{\kappa^2}{\omega} \sum_{m=1}^{\infty} \sum_{n=1}^{\infty} \begin{pmatrix} A_{mn}^{(M)} \exp\left(-jh_{mn}z_a\right) \\ B_{mn}^{(M)} \exp\left(jh_{mn}z_a\right) \end{pmatrix} \sin\frac{m\pi}{2}(\xi+1)\sin\frac{n\pi}{2}(\eta+1) \quad (7)$$

$$h_{mn} = \sqrt{\kappa^2 - (m\pi/2)^2 - p^2(n\pi/2)^2}, \quad \kappa = ka, \quad p = a/b(=1/q) \quad (8)$$

where $\xi = x/a$, $\eta = y/b$, and $z_a = z/a$ are the normalized coordinates.

For the diffracted waves in the half-spaces, we use the x- and y-components of the electric vector potential $\mathbf{F}$ (the use of the tangential components to the aperture is indispensable to incorporate the edge condition correctly), and they are given by Fourier spectral representations. The condition that the tangential electric field vanishes on the conducting screen can be satisfied by using the WS integrals [15], and we have

$$F_x^{d(i)} = (-1)^i a\epsilon \sum_{m=0}^{\infty}\sum_{n=0}^{\infty}\int_0^{\infty}\int_0^{\infty}\Big\{\Lambda_{2m}^{\sigma}(\alpha)\cos\alpha\xi\Big[A_{mn}^{x(i)}\Lambda_{2n}^{\tau}(\beta)\cos\beta\eta$$
$$+B_{mn}^{x(i)}\Lambda_{2n+1}^{\tau}(\beta)\sin\beta\eta\Big] + \Lambda_{2m+1}^{\sigma}(\alpha)\sin\alpha\xi\Big[C_{mn}^{x(i)}\Lambda_{2n}^{\tau}(\beta)\cos\beta\eta \tag{9}$$
$$+D_{mn}^{x(i)}\Lambda_{2n+1}^{\tau}(\beta)\sin\beta\eta\Big]\Big\}\frac{\exp\Big[(-1)^i\zeta(\alpha,\beta)z_a^{(i)}\Big]}{\zeta(\alpha,\beta)}d\alpha d\beta$$

$$F_y^{d(i)} = (-1)^i a\epsilon \sum_{m=0}^{\infty}\sum_{n=0}^{\infty}\int_0^{\infty}\int_0^{\infty}\Big\{\Lambda_{2m}^{\tau}(\alpha)\cos\alpha\xi\Big[A_{mn}^{y(i)}\Lambda_{2n}^{\sigma}(\beta)\cos\beta\eta$$
$$+B_{mn}^{y(i)}\Lambda_{2n+1}^{\sigma}(\beta)\sin\beta\eta\Big] + \Lambda_{2m+1}^{\tau}(\alpha)\sin\alpha\xi\Big[C_{mn}^{y(i)}\Lambda_{2n}^{\sigma}(\beta)\cos\beta\eta \tag{10}$$
$$+D_{mn}^{y(i)}\Lambda_{2n+1}^{\sigma}(\beta)\sin\beta\eta\Big]\Big\}\frac{\exp\Big[(-1)^i\zeta(\alpha,\beta)z_a^{(i)}\Big]}{\zeta(\alpha,\beta)}d\alpha d\beta$$

$$\zeta(\alpha,\beta) = \sqrt{\alpha^2+p^2\beta^2-\kappa^2},\quad z_a^{(1)} = z_a,\quad z_a^{(2)} = z_a + d_a,\quad d_a = d/a \tag{11}$$

where $\Lambda_{\ell}^{\nu}(x) = J_{\ell+\nu}(x)/x^{\nu}$ and $J_n(x)$ is the Bessel function of order n. The index i corresponds to the region number ($i = 1$ for region I and $i = 2$ for region II). The expressions of (9) and (10) are the Kobayashi potentials for the present problem. Parameters σ and τ are selected so as to incorporate the correct edge property into the electric field [8], and the correct values for the right-angled wedge, which are seemed to be valid for a deep hole, are $\sigma = 7/6$ and $\tau = 1/6$ $(= \sigma - 1)$. $A_{mn}^{x(i)} \sim D_{mn}^{x(i)}$ and $A_{mn}^{y(i)} \sim D_{mn}^{y(i)}$ are determined from the remaining boundary conditions on the aperture.

2.2 Matrix equation

We enforce the remaining boundary conditions that the tangential EM fields are continuous on the aperture and the resultant equations are projected into functional spaces of trigonometric functions and Gegenbauer polynomials by using their orthogonality (for details, see [15]). After some lengthy analysis, we have matrix equations for the expansion coefficients:

$$\begin{bmatrix} K_{Amnst}^{(u,v)} + S_{Amnst}^{\pm(u,v)} & (-1)^{u+v}p\Big\{G_{Amnst}^{(u,v)} + T_{Amnst}^{\pm(u,v)}\Big\} \\ (-1)^{u+v}q\Big\{G_{Bmnst}^{(\overline{u},\overline{v})} + T_{Bmnst}^{\pm(\overline{u},\overline{v})}\Big\} & K_{Bmnst}^{(\overline{u},\overline{v})} + S_{Bmnst}^{\pm(\overline{u},\overline{v})}\end{bmatrix}\begin{bmatrix} X_{mn}^{\pm(u,v)} \\ Y_{mn}^{\pm(\overline{u},\overline{v})}\end{bmatrix}$$
$$= 2j\begin{bmatrix} j^{u+v}P_x\Lambda_{2s+u}^{\tau'}(\kappa_a)\Lambda_{2t+v}^{\sigma'}(\kappa_b) \\ j^{\overline{u}+\overline{v}}q^2P_y\Lambda_{2s+\overline{u}}^{\sigma'}(\kappa_a)\Lambda_{2t+\overline{v}}^{\tau'}(\kappa_b)\end{bmatrix},\quad \begin{cases} s = 0,1,2,\cdots \\ t = 0,1,2,\cdots\end{cases} \tag{12}$$

$$(u,v) = (0,0),\ (0,1),\ (1,0),\ (1,1),\ \overline{u} = 1-u,\ \overline{v} = 1-v. \tag{13}$$

Here, $X_{mn}^{\pm(u,v)}$, $Y_{mn}^{\pm(u,v)}$, P_x, P_y, κ_a, and κ_b are defined as follows:

$$X_{mn}^{\pm(u,v)} = X_{mn}^{(1)(u,v)} \pm X_{mn}^{(2)(u,v)},\quad Y_{mn}^{\pm(u,v)} = Y_{mn}^{(1)(u,v)} \pm Y_{mn}^{(2)(u,v)} \tag{14}$$
$$X_{mn}^{(i)(0,0)} = A_{mn}^{x(i)},\ X_{mn}^{(i)(0,1)} = B_{mn}^{x(i)},\ X_{mn}^{(i)(1,0)} = C_{mn}^{x(i)},\ X_{mn}^{(i)(1,1)} = D_{mn}^{x(i)} \tag{15}$$
$$Y_{mn}^{(i)(0,0)} = A_{mn}^{y(i)},\ Y_{mn}^{(i)(0,1)} = B_{mn}^{y(i)},\ Y_{mn}^{(i)(1,0)} = C_{mn}^{y(i)},\ Y_{mn}^{(i)(1,1)} = D_{mn}^{y(i)} \tag{16}$$
$$P_x = \kappa(E_1\cos\theta_0\cos\phi_0 + E_2\sin\phi_0),\quad \kappa_a = \kappa\sin\theta_0\cos\phi_0 \tag{17}$$
$$P_y = \kappa(E_1\cos\theta_0\sin\phi_0 - E_2\cos\phi_0),\quad \kappa_b = q\kappa\sin\theta_0\sin\phi_0. \tag{18}$$

In (12), $K_{A,B}$ and $G_{A,B}$ are the double infinite integrals of four Bessel functions, and they are given by

$$K^{(u,v)}_{Amnst} = \int_0^\infty \int_0^\infty \frac{\kappa^2 - \alpha^2}{\zeta(\alpha,\beta)} \Lambda^{\sigma}_{2m+u}(\alpha)\Lambda^{\tau'}_{2s+u}(\alpha)\Lambda^{\tau}_{2n+v}(\beta)\Lambda^{\sigma'}_{2t+v}(\beta)d\alpha d\beta \tag{19}$$

$$K^{(u,v)}_{Bmnst} = \int_0^\infty \int_0^\infty \frac{q^2\kappa^2 - \beta^2}{\zeta(\alpha,\beta)} \Lambda^{\tau}_{2m+u}(\alpha)\Lambda^{\sigma'}_{2s+u}(\alpha)\Lambda^{\sigma}_{2n+v}(\beta)\Lambda^{\tau'}_{2t+v}(\beta)d\alpha d\beta \tag{20}$$

$$G^{(u,v)}_{Amnst} = \int_0^\infty \int_0^\infty \frac{\alpha\beta}{\zeta(\alpha,\beta)} \Lambda^{\tau}_{2m+1-u}(\alpha)\Lambda^{\tau'}_{2s+u}(\alpha)\Lambda^{\sigma}_{2n+1-v}(\beta)\Lambda^{\sigma'}_{2t+v}(\beta)d\alpha d\beta \tag{21}$$

$$G^{(u,v)}_{Bmnst} = \int_0^\infty \int_0^\infty \frac{\alpha\beta}{\zeta(\alpha,\beta)} \Lambda^{\sigma}_{2m+1-u}(\alpha)\Lambda^{\sigma'}_{2s+u}(\alpha)\Lambda^{\tau}_{2n+1-v}(\beta)\Lambda^{\tau'}_{2t+v}(\beta)d\alpha d\beta \tag{22}$$

and $S_{A,B}$ and $T_{A,B}$ are the double infinite series of four Bessel functions and they are given by

$$S^{\pm(u,v)}_{Amnst} = \pi^2 \sum_{m'=0}^{\infty}\sum_{n'=0}^{\infty} \frac{\Gamma^{\pm}_{2m'+1+u,2n'+v}}{1+\delta_{0,2n'+v}} \left[\kappa^2 - \left(\frac{2m'+1+u}{2}\pi\right)^2\right] \Lambda^{\sigma}_{2m+u}\left(\frac{2m'+1+u}{2}\pi\right)$$
$$\times \Lambda^{\tau'}_{2s+u}\left(\frac{2m'+1+u}{2}\pi\right)\Lambda^{\tau}_{2n+v}\left(\frac{2n'+v}{2}\pi\right)\Lambda^{\sigma'}_{2t+v}\left(\frac{2n'+v}{2}\pi\right) \tag{23}$$

$$S^{\pm(u,v)}_{Bmnst} = \pi^2 \sum_{m'=0}^{\infty}\sum_{n'=0}^{\infty} \frac{\Gamma^{\pm}_{2m'+u,2n'+1+v}}{1+\delta_{0,2m'+u}} \left[q^2\kappa^2 - \left(\frac{2n'+1+v}{2}\pi\right)^2\right] \Lambda^{\tau}_{2m+u}\left(\frac{2m'+u}{2}\pi\right)$$
$$\times \Lambda^{\sigma'}_{2s+u}\left(\frac{2m'+u}{2}\pi\right)\Lambda^{\sigma}_{2n+v}\left(\frac{2n'+1+v}{2}\pi\right)\Lambda^{\tau'}_{2t+v}\left(\frac{2n'+1+v}{2}\pi\right) \tag{24}$$

$$T^{\pm(u,v)}_{Amnst} = \pi^2 \sum_{m'=0}^{\infty}\sum_{n'=0}^{\infty} \Gamma^{\pm}_{2m'+1+u,2n'+v} \left(\frac{2m'+1+u}{2}\pi\right)\left(\frac{2n'+v}{2}\pi\right)\Lambda^{\tau}_{2m+1-u}\left(\frac{2m'+1+u}{2}\pi\right)$$
$$\times \Lambda^{\tau'}_{2s+u}\left(\frac{2m'+1+u}{2}\pi\right)\Lambda^{\sigma}_{2n+1-v}\left(\frac{2n'+v}{2}\pi\right)\Lambda^{\sigma'}_{2t+v}\left(\frac{2n'+v}{2}\pi\right) \tag{25}$$

$$T^{\pm(u,v)}_{Bmnst} = \pi^2 \sum_{m'=0}^{\infty}\sum_{n'=0}^{\infty} \Gamma^{\pm}_{2m'+u,2n'+1+v} \left(\frac{2m'+u}{2}\pi\right)\left(\frac{2n'+1+v}{2}\pi\right)\Lambda^{\sigma}_{2m+1-u}\left(\frac{2m'+u}{2}\pi\right)$$
$$\times \Lambda^{\sigma'}_{2s+u}\left(\frac{2m'+u}{2}\pi\right)\Lambda^{\tau}_{2n+1-v}\left(\frac{2n'+1+v}{2}\pi\right)\Lambda^{\tau'}_{2t+v}\left(\frac{2n'+1+v}{2}\pi\right) \tag{26}$$

where δ_{mn} is the Kronecker delta and

$$\Gamma^{\pm}_{mn} = \frac{1}{\gamma_{mn}} \frac{1 \mp \exp\left(-\gamma_{mn} d_a\right)}{1 \pm \exp\left(-\gamma_{mn} d_a\right)}, \quad \gamma_{mn} = \sqrt{\left(\frac{m\pi}{2}\right)^2 + p^2\left(\frac{n\pi}{2}\right)^2 - \kappa^2} = jh_{mn}. \tag{27}$$

In Eqs. (19)–(26), parameters σ' and τ' are determined by considering the edge property of the magnetic field [8]. When the hole is deep, we can use the edge condition of the right-angled wedge, and for this case the values of $\sigma' = -1/6$ $(= -\tau)$ and $\tau' = 5/6$ $(= \sigma' + 1)$ can be selected.

2.3 Physical quantities

The expression of the far-field is obtained by applying the stationary phase method to (9) and (10), and we have

$$F_x^{d(i)} = (-1)^i \frac{\pi q a^2 \epsilon}{2} \frac{\exp(-jkr)}{r} \sum_{m=0}^{\infty} \sum_{n=0}^{\infty} \left\{ \Lambda_{2m}^{\sigma}(\kappa_x) \left[A_{mn}^{x(i)} \Lambda_{2n}^{\tau}(\kappa_y) + jB_{mn}^{x(i)} \right.\right.$$
$$\left.\left. \times \Lambda_{2n+1}^{\tau}(\kappa_y) \right] + \Lambda_{2m+1}^{\sigma}(\kappa_x) \left[jC_{mn}^{x(i)} \Lambda_{2n}^{\tau}(\kappa_y) - D_{mn}^{x(i)} \Lambda_{2n+1}^{\tau}(\kappa_y) \right] \right\} \tag{28}$$

$$F_y^{d(i)} = (-1)^i \frac{\pi q a^2 \epsilon}{2} \frac{\exp(-jkr)}{r} \sum_{m=0}^{\infty} \sum_{n=0}^{\infty} \left\{ \Lambda_{2m}^{\tau}(\kappa_x) \left[A_{mn}^{y(i)} \Lambda_{2n}^{\sigma}(\kappa_y) + jB_{mn}^{y(i)} \right.\right.$$
$$\left.\left. \times \Lambda_{2n+1}^{\sigma}(\kappa_y) \right] + \Lambda_{2m+1}^{\tau}(\kappa_x) \left[jC_{mn}^{y(i)} \Lambda_{2n}^{\sigma}(\kappa_y) - D_{mn}^{y(i)} \Lambda_{2n+1}^{\sigma}(\kappa_y) \right] \right\} \tag{29}$$

$$\kappa_x = \kappa \sin\theta \cos\phi, \quad \kappa_y = q\kappa \sin\theta \sin\phi. \tag{30}$$

In the above expressions, θ and ϕ are the spherical coordinate angles (the angles of diffraction) shown in **Figure 1**, and r is the distance from the center point of each aperture to the observation point, that is, $r = \sqrt{x^2 + y^2 + z^2}$ for region I and $r = \sqrt{x^2 + y^2 + (z+d)^2}$ for region II (the origin of the spherical coordinate system for region II is selected at the center of the lower aperture). Far electromagnetic fields are obtained from $\mathbf{E} = -\epsilon^{-1}\nabla \times \mathbf{F}$ and $\mathbf{H} = Y\mathbf{i}_r \times \mathbf{E}$ ($\mathbf{i}_r$ is a unit vector of a radial direction). The transmission coefficient T is defined by the ratio of the total radiated power W_T into the $z < -d$ dark space to the power of the incident plane wave W_i on the aperture; $T = W_T/W_i$. W_T can be obtained by integrating the radiated power over the lower hemisphere, and we use the Gauss-Legendre quadrature in practical computation. W_i is analytically calculated, and the result is $W_i = 4abY\cos\theta_0$ for $|\mathbf{E}^{inc}| = 1$. For oblique incidence, $T\cos\theta_0$, which is the same as the transmission coefficient normalized by the normal incident power $W_i(\theta_0 = 0)$, is useful to evaluate the actual transmitted power into region II.

The expression of the aperture electric field is obtained by differentiating (9) and (10) with respect to z, and the resultant expression is given by

$$E_x^{d(i)} = \frac{\Gamma(\sigma)\Gamma(\tau)}{2^{2-(\sigma+\tau)}} (1-\xi^2)^{\tau-\frac{1}{2}} (1-\eta^2)^{\sigma-\frac{1}{2}} \sum_{m=0}^{\infty} \sum_{n=0}^{\infty} (-1)^{m+n} \left\{ \frac{\Gamma(2m+1)}{\Gamma(2m+2\tau)} C_{2m}^{(\tau)}(\xi) \right.$$
$$\times \left[A_{mn}^{y(i)} \frac{\Gamma(2n+1)}{\Gamma(2n+2\sigma)} C_{2n}^{(\sigma)}(\eta) + B_{mn}^{y(i)} \frac{\Gamma(2n+2)}{\Gamma(2n+2\sigma+1)} C_{2n+1}^{(\sigma)}(\eta) \right] + \frac{\Gamma(2m+2)}{\Gamma(2m+2\tau+1)}$$
$$\left. \times C_{2m+1}^{(\tau)}(\xi) \left[C_{mn}^{y(i)} \frac{\Gamma(2n+1)}{\Gamma(2n+2\sigma)} C_{2n}^{(\sigma)}(\eta) + D_{mn}^{y(i)} \frac{\Gamma(2n+2)}{\Gamma(2n+2\sigma+1)} C_{2n+1}^{(\sigma)}(\eta) \right] \right\} \tag{31}$$

$$E_y^{d(i)} = -\frac{\Gamma(\sigma)\Gamma(\tau)}{2^{2-(\sigma+\tau)}} (1-\xi^2)^{\sigma-\frac{1}{2}} (1-\eta^2)^{\tau-\frac{1}{2}} \sum_{m=0}^{\infty} \sum_{n=0}^{\infty} (-1)^{m+n} \left\{ \frac{\Gamma(2m+1)}{\Gamma(2m+2\sigma)} C_{2m}^{(\sigma)}(\xi) \right.$$
$$\times \left[A_{mn}^{x(i)} \frac{\Gamma(2n+1)}{\Gamma(2n+2\tau)} C_{2n}^{(\tau)}(\eta) + B_{mn}^{x(i)} \frac{\Gamma(2n+2)}{\Gamma(2n+2\tau+1)} C_{2n+1}^{(\tau)}(\eta) \right] + \frac{\Gamma(2m+2)}{\Gamma(2m+2\sigma+1)}$$
$$\left. \times C_{2m+1}^{(\sigma)}(\xi) \left[C_{mn}^{x(i)} \frac{\Gamma(2n+1)}{\Gamma(2n+2\tau)} C_{2n}^{(\tau)}(\eta) + D_{mn}^{x(i)} \frac{\Gamma(2n+2)}{\Gamma(2n+2\tau+1)} C_{2n+1}^{(\tau)}(\eta) \right] \right\} \tag{32}$$

where $C_{\ell}^{(\alpha)}(x)$ are the Gegenbauer polynomials and $\Gamma(x)$ is the gamma function.

3. Numerical results and discussion

To obtain the numerical results for the physical quantities of interest, the matrix Eq. (12) must be solved. The matrix elements consist of double infinite integrals and double infinite series that converge rather slowly, and we can compute them with the desired accuracy by applying the Hongo method [10]. For the double infinite integrals, the full range of integration that is equal to the first quadrant of the $(\alpha, p\beta)$ space is divided into several subdomains. The infinite integrals over the subdomains outside a very large radius are transformed into more simple forms by applying asymptotic expansions of the Bessel functions (the integrands of the infinite integrals are represented by some elementary functions), and the infinite integrals of the elementary functions are analytically carried out. The finite integrals of the Bessel functions are numerically performed by applying the Gauss-Legendre quadrature scheme (examples of the treatment are shown in [10–12]). For the double infinite series, the asymptotic expansions are applied to the summands (details are discussed in [13–15]). Once the expansion coefficients are determined by solving the matrix equations, we can readily obtain the physical quantities from their concise expressions derived in the previous section. Unless otherwise noted, the edge condition of the right-angled wedge ($\sigma = 7/6$, $\tau = 1/6$, $\sigma' = -1/6$, and $\tau' = 5/6$) is used, and $|\mathbf{E}^{inc}| = 1$ is selected in all of the following calculations. In practical computation, the matrix size is truncated to $2n_k^2 \times 2n_k^2$, where $n_k - 1$ is the maximum value of indices $m, n, s,$ and t in (12). By changing the value of n_k, we can numerically verify the convergence of the solution.

3.1 Convergence of the KP solution

We first calculated the transmission coefficient for various aperture sizes, aspect ratios, screen thicknesses, and incident angles. **Figure 2** shows examples of the transmission coefficient T as a function of the n_k value for square holes of $2a = \lambda/2$ ($ka = \pi/2$), $2a = \lambda$ ($ka = \pi$), and $2a = 3\lambda$ ($ka = 3\pi$) (λ is the wavelength) at $\theta_0 = 0$, 45 degrees ($\phi_0 = 0$). The subscripts E and H denote the E-polarization ($E_1 = 1$,

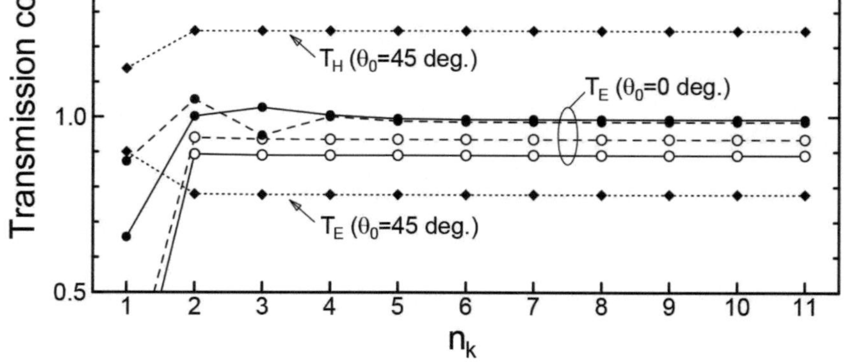

Figure 2.
Convergence of T_E and T_H for square apertures of $2a = \lambda/2$, λ, and 3λ ($ka = \pi/2$, π, and 3π) at $\theta_o = 0°$, $45°$, and $\phi_o = 0°$.

$E_2 = 0$) and H-polarization ($E_1 = 0, E_2 = 1$), respectively. For the normal incidence on the square hole, T_H is the same as T_E. From the figure, we see that the convergence is very rapid, and, for example, the results of $d/a = 0.5$ at $\theta_0 = \phi_0 = 0$ degrees have fully converged to four decimal places when $n_k = 4$ for $ka = \pi/2 - 10^{-10}$, $n_k = 5$ for $ka = \pi - 10^{-10}$, and $n_k = 7$ for $ka = 3\pi - 10^{-10}$. By the way, the summands of the series of (23)-(26) have singular points at $\gamma_{mn} = 0$ that arise for $ka = \ell\pi/2$ $(2a = \ell\lambda/2)$, $(\ell = 1, 2, 3, \ldots)$. We can remove this singularity appropriately in the numerical calculation, but we here avoided the singularity by slightly shifting the ka value by $\pm 10^{-10}$. The results for $ka = \pi/2 \pm 10^{-10}$ have completely agreed for eight decimal places, and similar situations had held for other aperture sizes related with the singular points. Thus, this treatment is also applied for calculating the other physical quantities.

The convergent property of the KP solution was also examined for other physical quantities. **Figures 3–5** show the far-field patterns of three kinds of screen thicknesses for various sizes of square apertures ($ka = \pi/2 - 10^{-10}$ (small), $2\pi - 10^{-10}$, and $5\pi - 10^{-10}$ (very large)) at the normal incidence. The patterns are computed with $P(\theta) = \lim_{r\to\infty} 4\pi r^2 \left|\mathbf{E}^{d(i)}\right|^2 / \left|\mathbf{E}^{inc}\right|^2$, and they are given for different n_k values. It can be seen from the figure that the plotted lines converge with relatively small value of n_k ($n_k = 2$ for $ka = \pi/2 - 10^{-10}$, $n_k = 4$ for $ka = 2\pi - 10^{-10}$, and $n_k = 9$ for $ka = 5\pi - 10^{-10}$). **Figures 6** and **7** show the amplitude distributions of aperture electric field E_y (the co-polarized component) for thin ($d/a = 0.001$) and thick ($d/a = 0.5$) screen cases, respectively. The rate of convergence of thick case is very fast but is very slow for thin case, and we think this is caused by using the edge condition of the right-angled wedge (in spite of the use of the right-angled wedge condition, the convergence of the far-field is very fast as shown in **Figure 3**. This is due to the fact that the far-field has a stationary nature, that is, it changes little for small changes in field distribution). We also calculate E_y for thin screen case by applying the edge condition of zero-thickness plate, and the results corresponding to **Figure 6** are given in **Figure 8**. Needless to say, the rate of convergence of the aperture field under the zero-thickness condition is much faster than that of the right-angled wedge condition, especially as the aperture is smaller.

3.2 Transmission properties

We next investigate the effect of the hole size and shape, screen thickness, and polarization on the transmission properties. Some results of the transmission coefficient are compared with those by other methods (Jin and Volakis [1], Brok and Urbach [7]). Unless otherwise noted, the following computations are made for $n_k = 6$. **Figure 9** shows the transmission coefficient T_E for square ($b/a = 1$) and rectangular ($b/a = 0.2$) holes less than the half-wavelength as a function of the normalized thickness d/a (in this case, all the waveguide modes become evanescent modes). It can be seen from the figure that the transmitted power rapidly decreases as the screen's thickness becomes large except for $ka = \pi/2 - 10^{-10}$, and this is due to the attenuation characteristic of the evanescent wave. However, rather strong transmission is observed for the elongated aperture ($b/a = 0.2$) when the value of d/a is small in spite of the fact that there is no propagating wave in the hole. This is considered to be closely related with the field singularity at long edges in addition to the small attenuation of the evanescent modes. By fitting asymptotic lines to T_E curves in **Figure 9** for the part where the value of d/a is large, we found an approximate expression for predicting the transmitted power, and it is given by

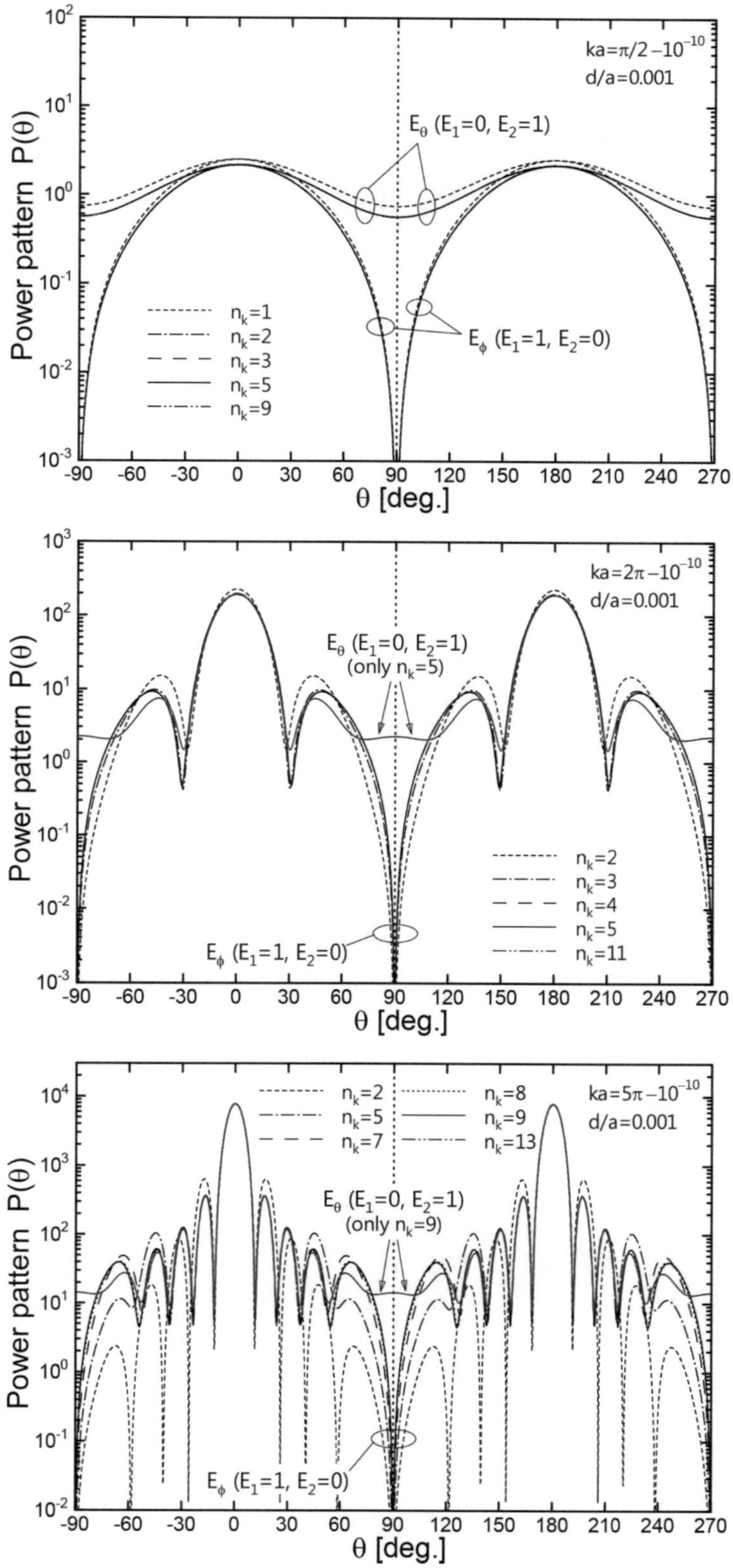

Figure 3.
Power patterns of the far-fields diffracted by a square aperture ($a = b$) for $d/a = 0.001$ (thin screen case) in the $\phi = 0°$ plane for the normal incidence ($\theta_o = \phi_o = 0°$).

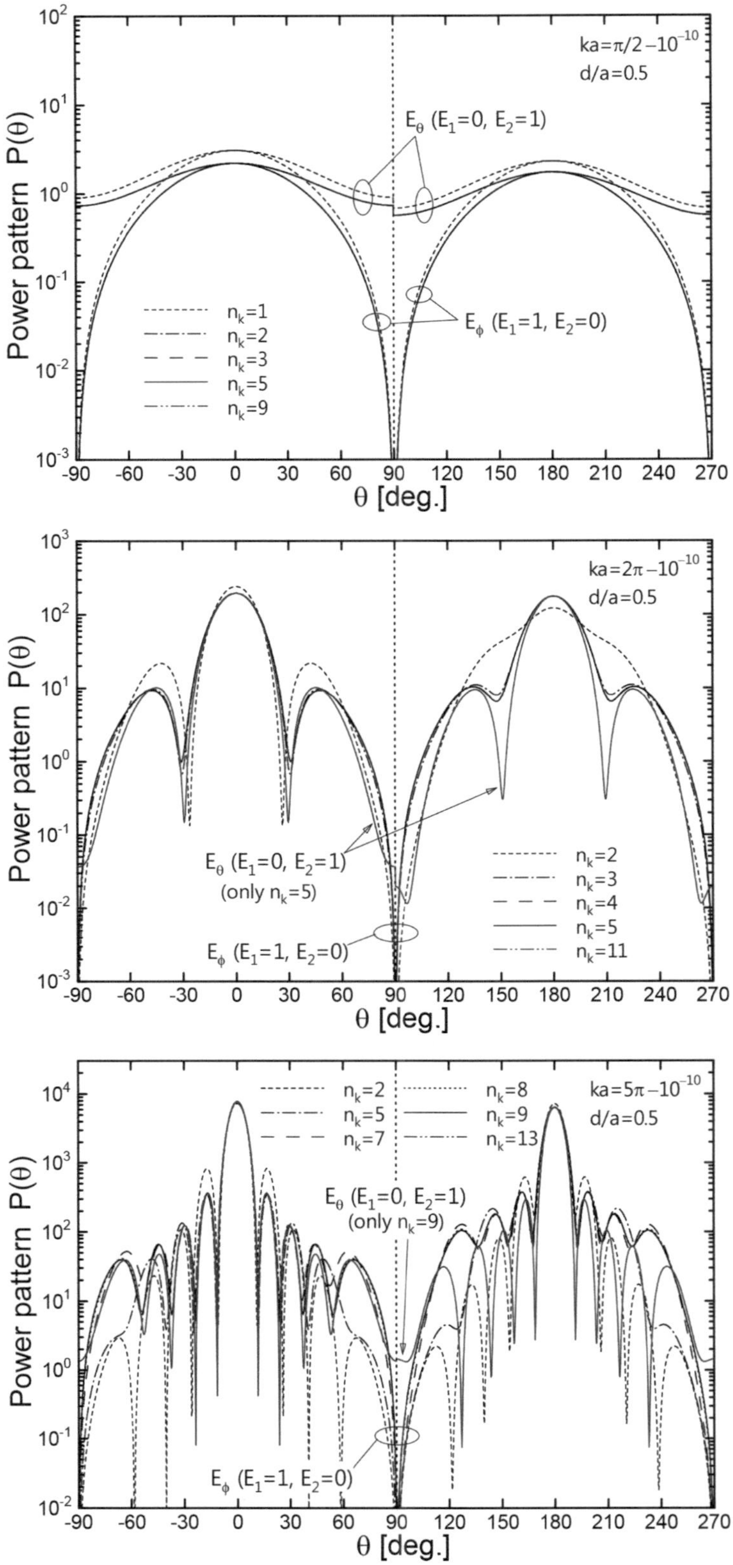

Figure 4.
Power patterns of the far-fields diffracted by a square aperture ($a = b$) for $d/a = 0.5$ (thick screen case) in the $\phi = 0°$ plane for the normal incidence ($\theta_o = \phi_o = 0°$).

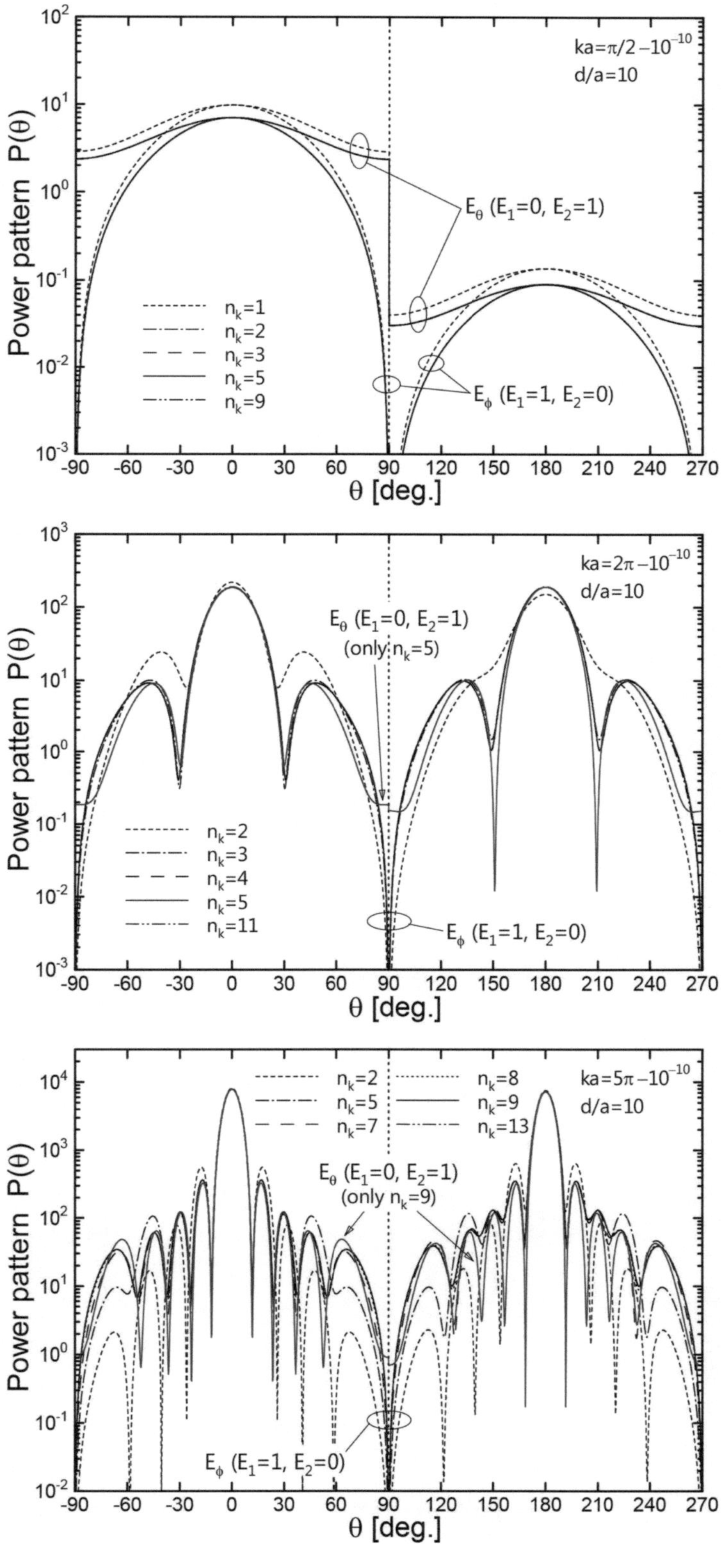

Figure 5.
Power patterns of the far-fields diffracted by a square aperture ($a = b$) for $d/a = 10$ (deep hole case) in the $\phi = 0°$ plane for the normal incidence ($\theta_o = \phi_o = 0°$).

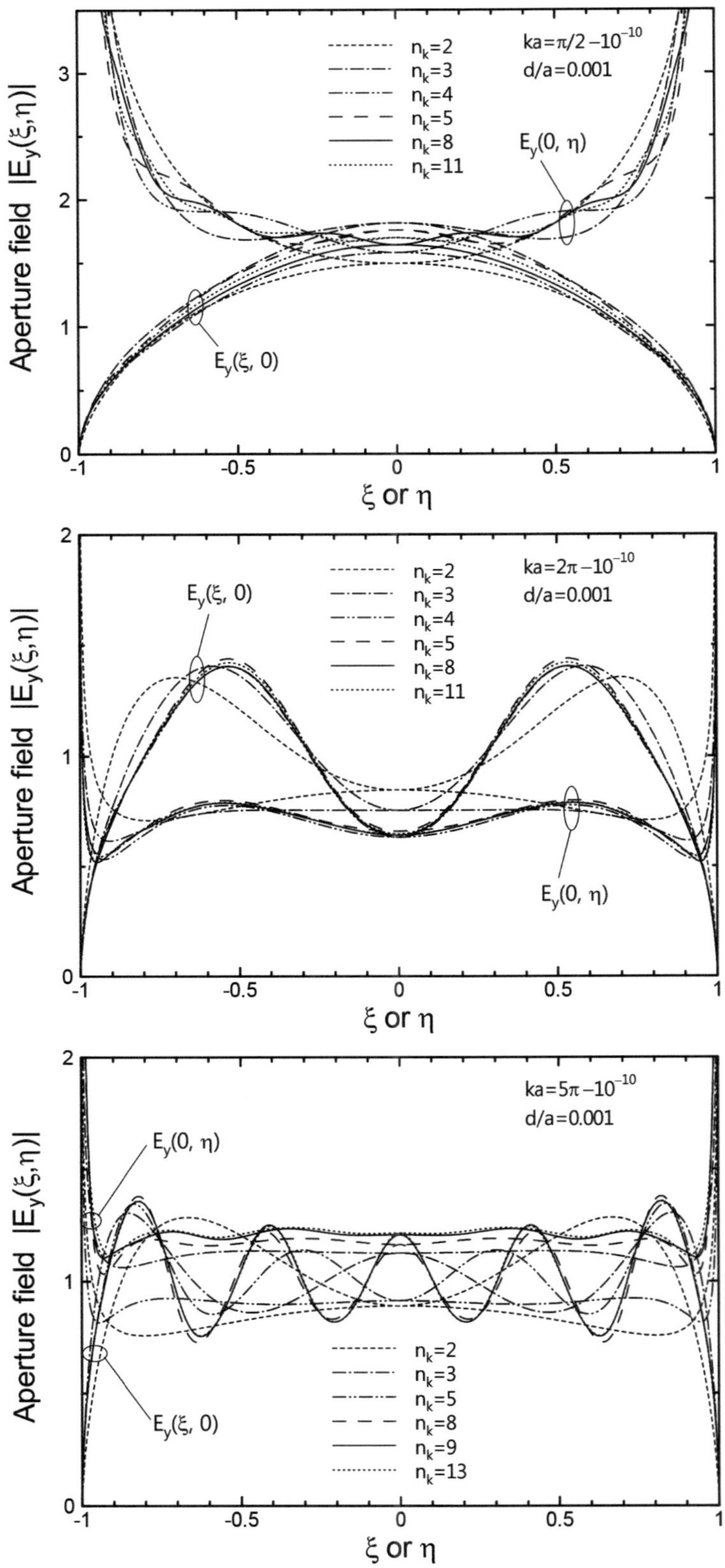

Figure 6.
Amplitude distribution of the y*-component of the aperture electric field* $|E_y(\xi,\eta)|$ *on the lower square aperture* $(a/b = 1, z = -d)$ *for* $d/a = 0.001$ *(thin screen case) at* $\theta_o = \phi_o = 0°$ *and E-polarized incidence.*

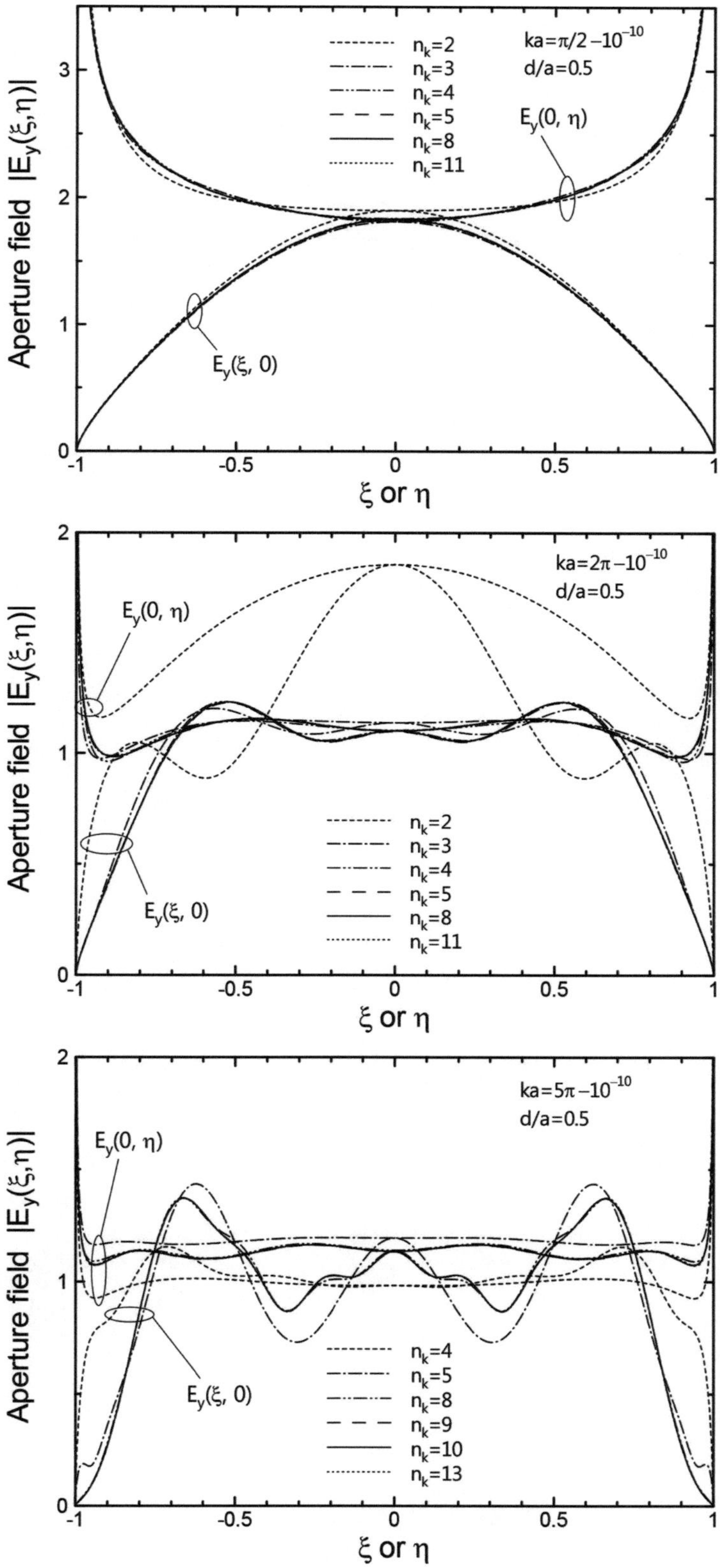

Figure 7.
Amplitude distribution of the y*-component of the aperture electric field* $|E_y(\xi, \eta)|$ *on the lower square aperture* $(a/b = 1, z = -d)$ *for* $d/a = 0.5$ *(thick screen case) at* $\theta_o = \phi_o = 0°$ *and E-polarized incidence.*

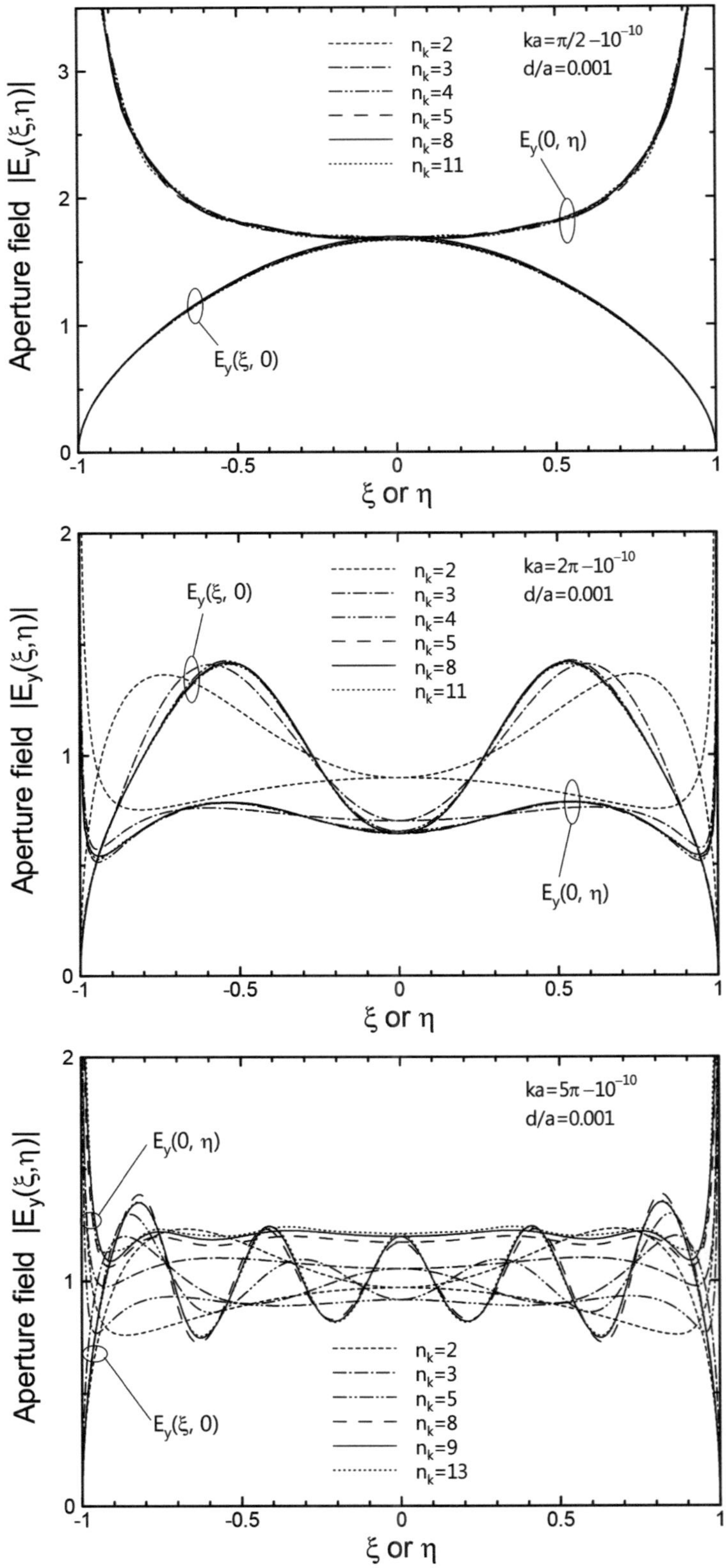

Figure 8.
Amplitude distribution of the y-component of the aperture electric field $|E_y(\xi,\eta)|$ on the lower square aperture ($a/b = 1$, $z = -d$) for $d/a = 0.001$ (thin screen case) at $\theta_o = \phi_o = 0°$ and E-polarized incidence. The edge condition of zero thickness screen ($\sigma = \tau' = 1$, $\tau = \sigma' = 0$) was used.

$$T_E = T_0 \exp\left(\frac{-2d}{a}\sqrt{\left(\frac{\pi}{2}\right)^2 - (ka)^2}\right) \quad (33)$$

where T_0 is a coefficient depending on ka and is shown in **Figure 10**.

The transmission coefficient is also calculated as a function of the normalized thickness d/a for various aperture sizes around the half-wavelength ($ka \approx \pi/2$), and the results are shown in **Figures 11** and **12** (our results completely agree with those by Brok and Urbach [7]). It can be seen from the figures that the T_E curves for apertures larger than the half-wavelength have oscillation, and this is due to the resonance of the

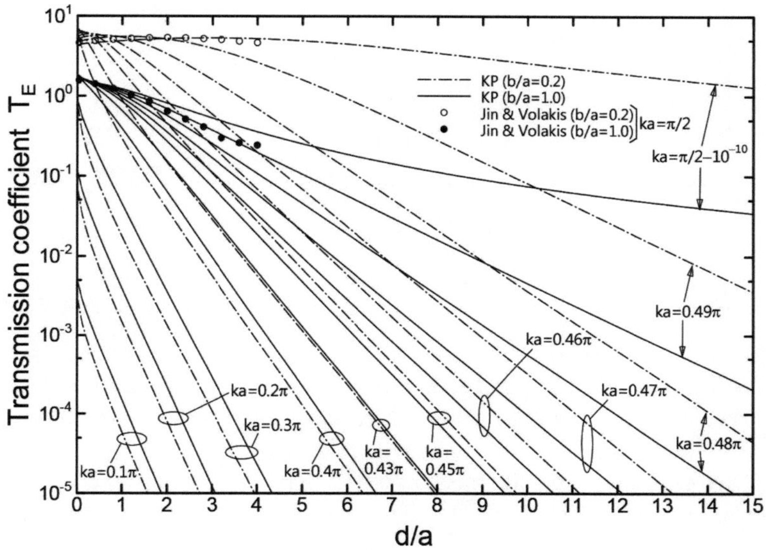

Figure 9.
T_E *as a function of the normalized screen thickness* d/a *for various ka values* ($b/a = 1, 0.2$) *at* $\theta_o = \phi_o = 0°$.

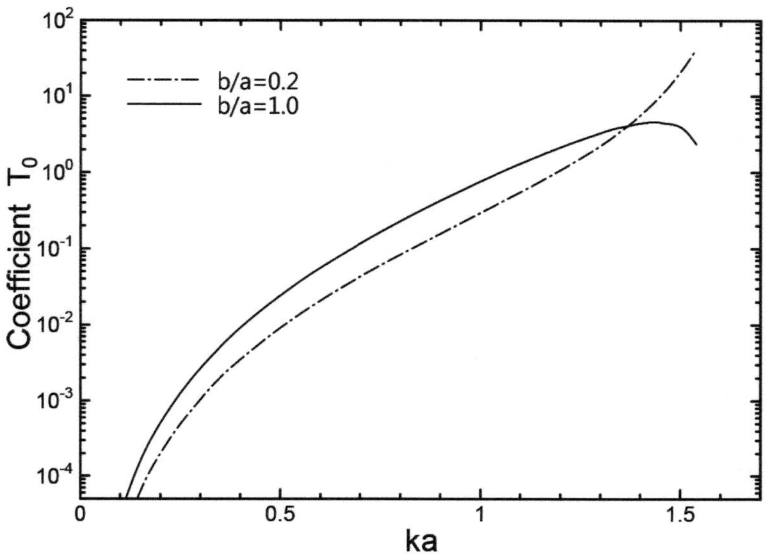

Figure 10.
Coefficient T_o *of asymptotic fitting lines for square* ($b/a = 1$) *and elongated rectangular* ($b/a = 0.2$) *apertures.*

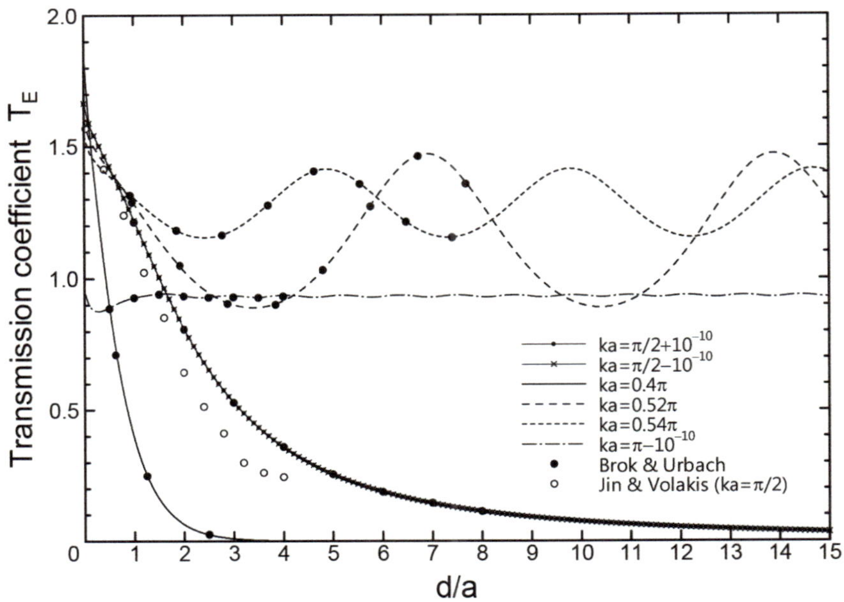

Figure 11.
T_E as a function of d/a for various ka values ($b/a = 1$) at $\theta_o = \phi_o = 0°$ and E-polarized incidence.

propagating waveguide modes. The results of the total transmitted power normalized by $W_i(\theta_0 = 0°)$ are shown in **Figures 13** and **14** for various aperture sizes and shapes, and the curves of the transmitted power of the large aperture are complex (we think this is also due to the resonance of the waveguide modes).

Finally, we calculate the transmission coefficient T_E as functions of the normalized aperture length $2a/\lambda$ for a constant b value ($2b = 0.1\lambda$) and of the aspect ratio b/a for a constant hole area ($2a \times 2b = (\pi/4)^2$), and the results are given in

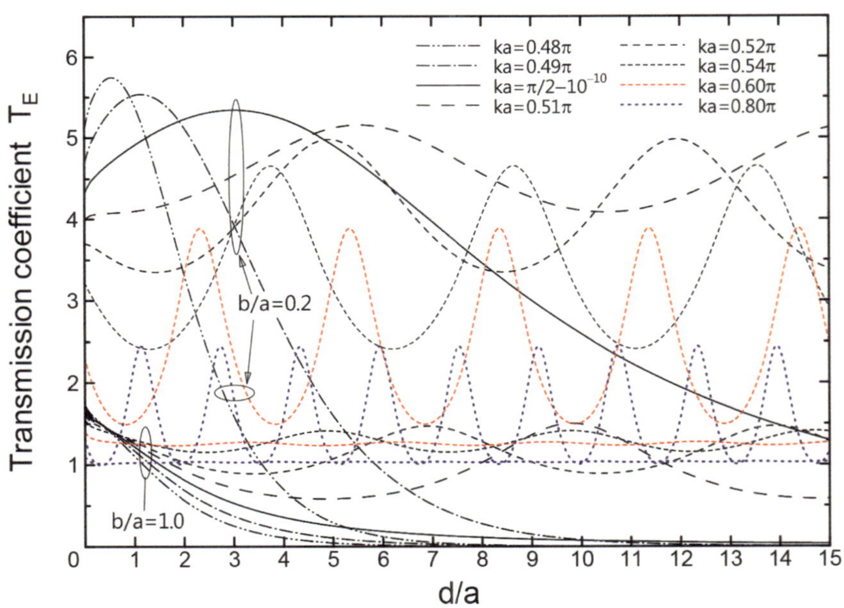

Figure 12.
Transmission coefficient T_E as a function of the normalized screen thickness d/a for various ka values ($b/a = 1$, 0.2) at $\theta_o = \phi_o = 0°$ and E-polarized incidence.

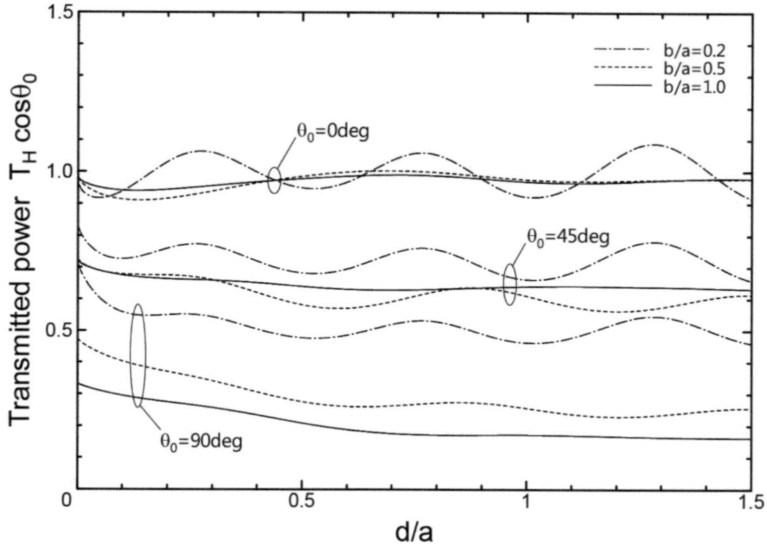

Figure 13.
Normalized transmitted power ($T_H \cos\theta_o$) through square and rectangular apertures ($ka = 2\pi - 10^{-10}$, $b/a = 1, 0.5, 0.2$) at $\theta_o = 0°, 45°, 90°$, and $\phi_o = 90°$.

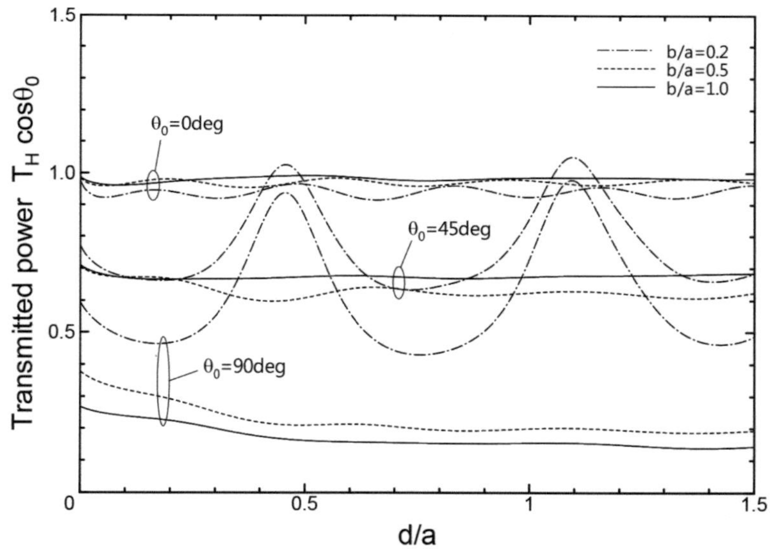

Figure 14.
Normalized transmitted power ($T_H \cos\theta_o$) through square and rectangular apertures ($ka = 3\pi - 10^{-10}$, $b/a = 1, 0.5, 0.2$) at $\theta_o = 0°, 45°, 90°$, and $\phi_o = 90°$.

Figures 15 and **16**, respectively. In obtaining these results, the value of n_k was chosen to be 5. In **Figure 15**, all the waveguide modes for $ka < \pi/2$ become evanescent waves, but the strong power transmission is observed, and the position of the peak point approaches to $ka = \pi/2$ (cutoff frequency) as the screen thickness becomes large. For $ka > \pi/2$, we can find that the T_E curves have oscillations that are caused by the resonance of the lowest propagating mode in the waveguide. In **Figure 16**, $b/a = 0.25$ is a value related with the cutoff frequency, and all the waveguide modes become evanescent waves for $b/a > 0.25$. The similar situations of **Figure 15** can be also seen in this case.

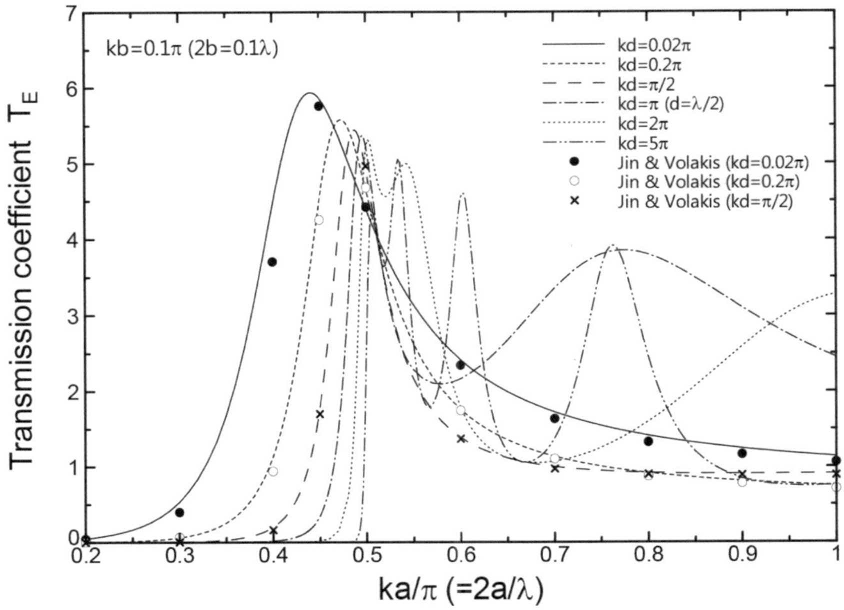

Figure 15.
T_E as a function of the normalized aperture length ka/π with $kb = 0.1\pi$ for various kd values at $\theta_o = \phi_o = 0°$.

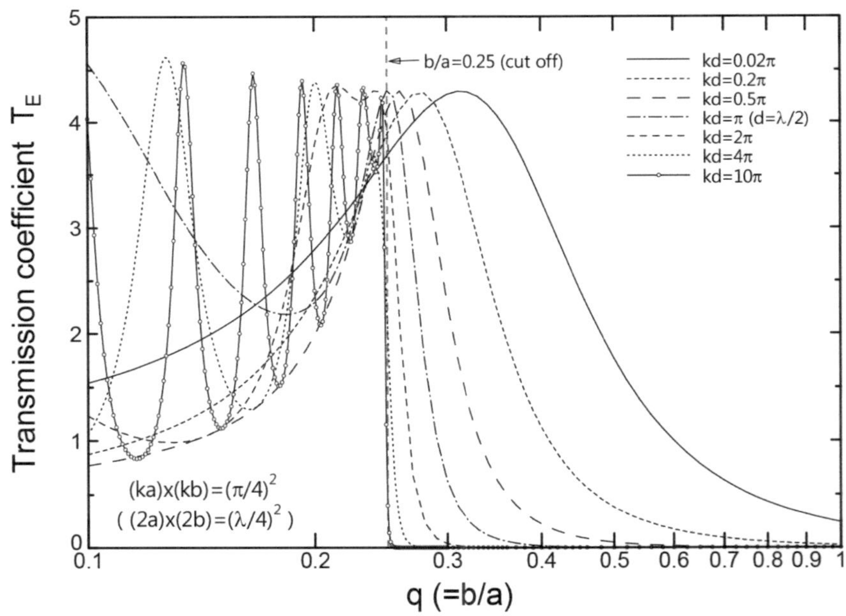

Figure 16.
T_E as a function of the aspect ratio $q = b/a$ with the same area ($ka \times kb = (\pi/4)^2$) for various kd values at $\theta_o = \phi_o = 0°$.

4. Conclusions

We have rigorously studied the diffraction of an electromagnetic plane wave by a rectangular hole in a perfectly conducting screen with a finite thickness by applying the KP method. The transmission coefficient, far-field pattern, and aperture electric field distribution were calculated to show the convergence of the solution.

The numerical results of the transmission coefficient were also presented as functions of the normalized thickness d/a to show the EM transmission property of the various sized holes. It was found from the results that the convergence of the KP solution is very rapid (for thin screen case, the edge condition of the zero-thickness plate is effective) and the transmitted power is strongly affected by the aperture size and shape, screen's thickness, incident angles, and polarization.

Acknowledgements

This work was partly supported by JSPS KAKENHI Grant Number 25390159 and 17K05150.

Author details

Hirohide Serizawa
National Institute of Technology, Numazu College, Numazu, Japan

*Address all correspondence to: serizawa@numazu-ct.ac.jp

References

[1] Jin J-M, Volakis JL. Electromagnetic scattering by and transmission through a three-dimensional slot in a thick conducting plane. IEEE Transactions on Antennas and Propagation. 1991;**39**(4): 543-550. DOI: 10.1109/8.81469

[2] Park HH, Eom HJ. Electromagnetic scattering from multiple rectangular apertures in a thick conducting screen. IEEE Transactions on Antennas and Propagation. 1999;**47**(6):1056-1060. DOI: 10.1109/8.777131

[3] Stevanovic I, Mosig JR. Efficient electromagnetic analysis of line-fed aperture antennas in thick conducting screens. IEEE Transactions on Antennas and Propagation. 2004;**52**(11):2896-2903. DOI: 10.1109/TAP.2004.835268

[4] Garcia-Vidal FJ, Martin-Moreno L, Ebbesen TW, Kuipers L. Light passing through subwavelength apertures. Reviews of Modern Physics. 2010;**82**(1): 729-787. DOI: 10.1103/RevModPhys. 82.729

[5] Garcia-Vidal FJ, Moreno E, Porto JA, Martin-Moreno L. Transmission of light through a single rectangular hole. Physical Review Letters. 2005;**95**(10): 103901. DOI: 10.1103/PhysRevLett. 95.103901

[6] Garcia-Vidal FJ, Martin-Moreno L, Moreno E, Kumar LKS, Gordon R. Transmission of light through a single rectangular hole in a real metal. Physical Review B. 2006;**74**(15):153411. DOI: 10.1103/PhysRevB.74.153411

[7] Brok JM, Urbach HP. Extraordinary transmission through 1, 2 and 3 holes in a perfect conductor, modelled by a mode expansion technique. Optics Express. 2006;**14**(7):2552-2572. DOI: 10.1364/OE.14.002552

[8] Meixner J. The behavior of electromagnetic fields at edges. IEEE Transactions on Antennas and Propagation. 1972;**AP-20**(4):442-446. DOI: 10.1109/TAP.1972.1140243

[9] Hongo K, Serizawa H. Kobayashi potential in electromagnetism. IEICE Transactions on Electronics. 2012;**E95-C** (1):3-9. DOI: 10.1587/transele.E95.C.3

[10] Hongo K, Sugaya H. Diffraction of an acoustic plane wave by a rectangular plate. Journal of Applied Physics. 1997; **82**(6):2719-2729. DOI: 10.1063/1.366266

[11] Hongo K, Serizawa H. Diffraction of electromagnetic plane wave by a rectangular plate and a rectangular hole in the conducting plate. IEEE Transactions on Antennas and Propagation. 1999;**47**(6):1029-1041. DOI: 10.1109/8.777128

[12] Hongo K, Serizawa H. Diffraction of an acoustic plane wave by a rectangular hole in an infinitely large rigid screen. The Journal of the Acoustical Society of America. 1999;**106**(1):29-35. DOI: 10.1121/1.427033

[13] Serizawa H, Hongo K. Evaluation of an acoustic plane wave transmitted through a rectangular hole in a thick hard screen. Wave Motion. 2002;**36**(2): 103-117. DOI: 10.1016/S0165-2125(01) 00116-0

[14] Serizawa H, Hongo K. Radiation from a flanged rectangular waveguide. IEEE Transactions on Antennas and Propagation. 2005;**53**(12):3953-3962. DOI: 10.1109/TAP.2005.859748

[15] Serizawa H, Hongo K. Transmission through a rectangular hole in a thick conducting screen. IEICE Technical Report (AP). 2009;**109**(218):31-36

[16] Serizawa H. Diffraction of a plane wave by a rectangular hole in a thick conducting screen. In: Proceedings of

the 9th European Conference on Antennas and Propagation (EuCAP 2015), MA12 EMTheory. Lisbon: IEEE; 2015

[17] Serizawa H. Plane wave diffraction by a small rectangular aperture in a thick conducting screen. In: Proceedings of the 2016 International Conference on Electromagnetics in Advanced Applications (ICEAA '16). Cairns: IEEE; 2016. pp. 890-893. DOI: 10.1109/ICEAA.2016.7731545

[18] Serizawa H. Plane wave diffraction by a large rectangular aperture in a thick conducting screen. In: Proceedings of the 2017 International Conference on Electromagnetics in Advanced Applications (ICEAA '17). Verona: IEEE; 2017. pp. 1612-1615. DOI: 10.1109/ICEAA.2017.8065597

Chapter 7

Nanoplasma Formation From Atomic Clusters Irradiated by Intense Femtosecond Lasers

Boucerredj Noureddine and Khaled Beggas

Abstract

The nanoplasma formations of Na and Kr clusters, which contained 2×10^7 atoms per cluster, irradiated by intense femtosecond laser field have been predicted in detail within the framework of the modified nanoplasma model. Based on this modified model, ionization process, heating, expansion, and explosion of the cluster have been studied. When laser intensity is above a critical value, it blows off all of electrons from the cluster and forms a non-neutral ion cloud. The analytic calculation provides time evolution of radius of the cluster, internal and external fields, coulomb and hydrodynamic pressures, electron density, and ion and electron energy. During the coulomb explosion of the resulting highly ionized, high-temperature nanoplasma, ions acquire their energy. It is shown that ultrafast ions are produced in this comparative study (4.4 keV for Kr cluster and 2.2 keV for Na cluster), which can be the source of energetic ions. We have found that the coulomb pressure is little than the hydrodynamic pressure for both clusters.

Keywords: large clusters, nanoplasma, laser plasma interaction, laser-produced plasma, ion source, cluster explosion

1. Introduction

During the last years, nanoplasma (nanometer scale plasma) formed from atomic clusters irradiated by high-power femtosecond lasers, capable of achieving a light intensity up to 10^{21}–10^{22} W/cm^2, has attracted a great attention and has been studied by several groups [1–5]. This large interest is due to the high energetic ions and electrons produced when rare gas clusters explode [5, 6]. The intense femtosecond laser field ionize atoms of the cluster several times and ions reaches a high degree of ionization [7–9], and are very effective to produce strong X-ray radiation [10]. In addition, fusion reaction has been observed in laser deuterium cluster interaction indicating that nuclear fusion occurs after explosion of the cluster [11]. The measurements of the energy absorption by the formed plasma have shown that the plasma absorbs a large fraction of the incident laser energy [12]. The interaction of rare gas and metallic clusters with ultrafast lasers has become a topic of great interest; a new fundamental insight into ultrafast laser-driven excitation and decay dynamics of many particle systems can be gained [13]. Macroscopic and hydrodynamic impact energy transfer from nanoplasma electrons to ions in exploding cluster using molecular dynamics simulation has been studied recently [14]. When the laser interacts with clusters, it changes the regime of electromagnetic wave propagation, and very efficient absorption of radiation has been demonstrated with a formation of very high temperature under dense plasma [15, 16]. The coulomb explosion plays an important role in the case of small-diameter clusters [17].

The interaction of intense femtosecond laser field with atomic cluster leads to the ionization of the parent atoms of the cluster. Ionization of the cluster begins on the rising edge of the laser field. Parent atoms absorb laser intensity and liberate a small number of electrons by direct optical ionization which produces a gas of electrons that forms the nanoplasma. The collisional ionization becomes important between electrons and ions leading to the rapid heating of the cluster to a non-equilibrium superheated state then the expansion of the cluster [18, 19].

The interaction of single clusters with intense femtosecond laser produces the hot electrons with energy up to 3 keV [20, 21]. After heating up the cluster, charge separation of the hot electrons and the repulsive coulomb force between the positive ions leads to the explosion of the cluster, and ions acquire their energy by the conversion of the energy deposited by the laser in the cluster to ion kinetic energy.

In this present research, we study the formation of the nanoscale plasma from the irradiated clusters by an intense femtosecond laser field, and we investigate a comparative behavior of the different physical parameters of the formed nanoplasma from rare gas (Kr) and metallic (Na) clusters within the framework of the modified nanoplasma model. After a general introduction on intense laser cluster interaction and plasma formation, we show in Section 2 a brief description of the nanoplasma model, followed by a discussion of the obtained results in Section 3. Finally, we report the obtained results in the conclusion section.

2. Nanoplasma model

The development of the numerical model called nanoplasma model was made firstly by Ditmire et al. [9], and after that reformulated by Megi et al. [18] by the addition of the term of electron collisions with surface in the expression of the electron-ion collision frequency. This numerical model has proved its success in reproducing experimental results of the interaction of single cluster with intense femtosecond laser field [19] and describes the interaction in terms of the formation of high-density nanoplasma.

In this model, the expanding cluster was treated as a spherical nanoplasma, subjected to the intense laser field. Ionization process such as field ionization is described by the Ammosov, Delone, and Krainov model [22]. Tunnel and collisional ionization are described using the Lotz formula [23]. Collisional ionization in the nanoplasma leads to the production of highly charged ions.

2.1 Ionization mechanisms

The basic mechanisms in this model are that, the laser strips electrons from the parent atoms by direct optical ionization and collisional ionization. The direct optical ionization begins when laser pulse liberates a small number of electrons. Field ionization is described in this model by Ammosov et al. [18, 22, 24], and collisional ionization, which occurs from inelastic collisions between electrons and ions, is described by the Lotz formula [9, 18, 23]. Due to the high excitation and hot electrons and densities reached inside the cluster, collisional ionization is the dominant ionization mechanism in the nanoplasma leading to the production of highly charged ions. The laser deposits its energy into the free electrons inside the cluster. These electrons absorb the magnetic power and heating up the ions. The heating in the cluster occurs through inverse bremsstrahlung collisions. The density of the energy deposition $\frac{\partial u}{\partial t}$ in the cluster is given by [9, 25]:

$$\frac{\partial u}{\partial t} = \varepsilon E_{\text{int}} \frac{\partial E_{\text{int}}}{\partial t} \tag{1}$$

where ε is the plasma dielectric constant.

In this model there are three main assumptions: First, we assume that $\lambda >> R_0$, where λ and R_0 are the laser wavelength and the initial cluster radius, respectively. Second, the electronic and ionic densities are distributed uniformly within the cluster. The third assumption is that the electrons are assumed to be instantaneously thermalized and their energy distribution is Maxwellian. The nanoplasma model takes into account the highly efficient collisional heating which is calculated in this model as the heating of a uniform dielectric sphere under the electric field of the laser. The internal electric field is given by [26].

$$E_{\text{int}}(t) = \frac{3}{|\varepsilon + 2|} E_{ext}(t), \tag{2}$$

where E_{ext} is the external electric field, t is the time, and ε is the plasma dielectric constant given by a Drude model,

$$\varepsilon = 1 - \frac{w_p^2}{w(w + i\nu)}, \tag{3}$$

where w is the plasma frequency, ν is the total electron-ion collision frequency, and w_p is the electronic plasma frequency given by

$$w_p = \sqrt{\frac{n_e e^2}{\varepsilon_0 m}}, \tag{4}$$

where n_e is the electron density, ε_0 is the vacuum dielectric constant, e is the electron charge, and m is the electron mass.

The external electric field is given by

$$\vec{E}_{ext}(t) = E_0 \sin(wt) f(t) \vec{e}_z,$$
$$\vec{E}_{ext}(t) = E_0 \exp\left(-2\ln 2\left(\frac{t}{\tau}\right)^2\right)\vec{e}_z,$$

where τ is the laser pulse duration at the full width at half maximum (FWHM), f(t) is a Gaussian distribution of laser profile, and E_0 is the amplitude of the laser field given as the maximum laser field, $E_{\max} = \sqrt{\frac{2I_{\max}}{\varepsilon_0 c}}$, where c is the light velocity.

The total electron-ion collision frequency including the electron-surface collision frequency, $\nu_s = v/R$, is given by the formula

$$\nu = \nu_{ei} + \nu_s = \nu_{ei} + \frac{v}{R}, \tag{5}$$

where v is the velocity of the electrons, R is the cluster radius, and ν_{ei} is the electron-ion collision frequency given by [27, 28]

$$\nu_{ei} = \frac{n_i \langle Z^2 \rangle e^4 \ln \Lambda_1}{2(\pi \varepsilon_0 m_e)^2 v^3}, \tag{6}$$

$$v = \sqrt{v_{th}^2 + v_{osc}^2}, \tag{7}$$

where $v_{th} = \sqrt{k_B T_e / m}$ is the thermal velocity and $v_{osc} = \frac{eE_{\text{int}}}{m_e \omega}$ is the oscillation velocity of the electron in the field, $<Z>$ is the mean ion charge, and $\ln\Lambda_1$ is the modified coulomb logarithm given by the expression.

$$\ln \Lambda_1 = \begin{cases} \frac{1}{4} \ln^2\left(\frac{1 + m_e v_{osc}^2}{\hbar w}\right), \text{for } \hbar w \gg k_B T_e \\ \frac{1}{4} \ln^2\left(1 + \frac{v_{osc}^2}{v_{th}^2}\right) + \ln\left(\frac{v_{osc}^2}{v_{th}^2} + \exp\left(\frac{1}{3}\sqrt{\frac{\pi}{2}}\right)\right) \ln \Lambda, \text{for } \hbar w \ll k_B T_e. \end{cases} \tag{8}$$

where k_B is the Boltzmann constant, T_e is the electron temperature (see [9] and [18]), and $\ln\Lambda$ is the classical coulomb logarithm [29].

2.2 Cluster expansion

The expansion of the nanoplasma is driven by two mechanisms. First, the hydrodynamic heating is driven by the hot electron pressure in the cluster. The heated electrons or the heated nanoplasma start to expand, then pull the heavy ions. The electron or hydrodynamic pressure has the expression

$$P_h = n_e k_B T_e, \tag{9}$$

where n_e is the electron density in the cluster $n_e = \frac{N_e}{\frac{4}{3}\pi R^3}$ and N_e is the number of electrons in the cluster.

Second, the coulomb explosion is driven by charge buildup on the cluster. After ionization, some hot electrons escape the coulomb barrier formed by ions, leaving a net positive charge on the cluster. The resulting repulsive coulomb force leads to the expansion of the cluster. The coulomb pressure is given by

$$P_{Coul} = \frac{3Q^2 e^2}{32\pi^2 \varepsilon_0 R^4}, \tag{10}$$

where Q is the built-up charge on the cluster due to electron escape (see [9] and [18]). The total pressure is given by

$$P = P_h + P_{Coul} \tag{11}$$

The coulomb pressure scales as $(1/R^4)$ shows that it will be important for small clusters. The hydrodynamic pressure scales as $(1/R^3)$ is therefore more important for large clusters. Then, the cluster radius equation is given by

$$\frac{\partial^2 R}{\partial t^2} = 5 \frac{P}{n_i m_i} \frac{1}{R}, \tag{12}$$

where n_i and m_i are the density and mass of ions in the cluster, respectively.

The ion energies from cluster explosions are due to the electron-ion collisions heating up the electrons, but not the ions. This can be seen from the electron-ion equilibration time. In the other hand, there is insufficient time for electron energy to be transferred to the ions through collisions. Instead, the ions gain energy in the hydrodynamic expansion, where the thermal energy of the electrons is converted to ion kinetic energy or through coulomb explosion. For a hydrodynamic expansion, the mean ion energy will be of order [24]

$$\langle E_{ion}\rangle \approx \langle Z\rangle k_B T_e, \tag{13}$$

where <Z> is the mean ion charge state. The coulomb explosion is the main process by which ions gain kinetic energy in the intense laser field.

The rate of the temperature decrease from the cluster expansion is given by [9]:

$$\left.\frac{\partial T_e}{\partial t}\right|_{\exp} = -2\frac{T_e}{R}\frac{\partial R}{\partial t} \tag{14}$$

2.3 Electron-ion thermalization

The transfer of thermal energy to the cloud ions in the cluster nanoplasma results from the coulomb collision of energetic electrons, which can be described by a thermal equilibration rate given by [9]

$$\left.\frac{\partial T_e}{\partial t}\right|_{eq} = -\frac{T_e - T_i}{\tau_{eq}} \tag{15}$$

The electron-ion equilibration time τ_{eq} is given by [9, 30]

$$\tau_{eq} = \frac{3m_e m_i}{8\sqrt{2\pi}n_i z^2 e^4 \ln\Lambda}\left(\frac{k_B T_e}{m_e} + \frac{k_B T_i}{m_i}\right)^{\frac{3}{2}} \tag{16}$$

where T_e is the electron temperature, T_i ion temperature, and m_e and m_i are the electron and ion mass, respectively.

3. Results and discussion

The calculation was carried out with an intense femtosecond laser including a peak intensity of 10^{17} W/cm^2, wave length of 390 nm, and pulse duration of

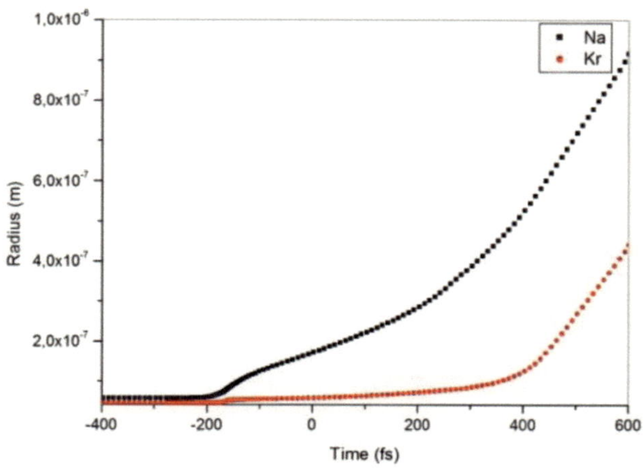

Figure 1.
Calculated time evolution of the cluster radius R of the Na and Kr clusters contained 2×10^7 atoms in cluster irradiated by an intense femtosecond laser with a pulse duration of 200 fs (FWHM), wavelength of 390 nm, and peak intensity of 10^{17} W/cm^2.

τ = 200 fs (full width at half maximum (FWHM)) irradiating the Kr and Na clusters containing 2×10^7 atoms per cluster. Inside the cluster, the temperature, the density gradient, and the internal field are assumed to be uniform. We have used the modified nanoplasma model to study the temporal variation of the different physical parameters (radius, electron temperature, expansion velocity, electron density, coulomb and hydrodynamic pressures, ion and electron energy, etc.) of the formed nanoplasma.

Time dependence of the cluster radius R for the rare gas Kr and metallic Na clusters is illustrated in **Figure 1**. Time evolution shows clearly a rapid expansion of the metallic cluster (Na) than the rare gas cluster (Kr). In the case of the metallic cluster, it expands rapidly due to the rapid ionization of the cluster.

The time zero in the calculation is the peak intensity of the laser at around −255fs for Na cluster and −230fs for Kr cluster; a small number of electrons are created through optical ionization; these electrons form the nanoplasma. The inelastic collisions between electrons and ions increase the number of electrons and ions. When electron density rises to reach $3n_{crit}$, where n_{crit} is the critical electron density given by the formula

$$n_{crit} = \frac{\varepsilon_0 m_e w^2}{e^2}, \tag{17}$$

where ε_0 is the vacuum permittivity and ω is the laser frequency.

The internal field is amplified and becomes greater than the external one (**Figure 2**), and we have a strong absorption of the laser energy by the cluster. The first resonance (between laser field and the formed nanoplasma) occurs at $t \approx -250$ fs, in the Na cluster, but at −230 fs it occurs in Kr cluster (**Figure 3**); the cluster expansion velocity increases dramatically; and more electrons are liberated through tunnel, laser-driven, and thermal ionization. The hot electrons can leave the cluster to the $3n_{crit}$ resonance.

When the electron density n_e is greater than $3n_{crit}$, the field inside the cluster is smaller than the external one. The maximum value of the ration $\frac{n_e}{n_{crit}}$ is about 3.5 for Na cluster than 7 for Kr cluster; this difference in the time of resonances and the maximum values of the ratio $\frac{n_e}{n_{crit}}$ is caused by the number of electrons in such cluster. When the hot electrons leave the cluster and the cluster expands, the electron density starts to decrease at time −176 fs for Na cluster and −163 fs for Kr

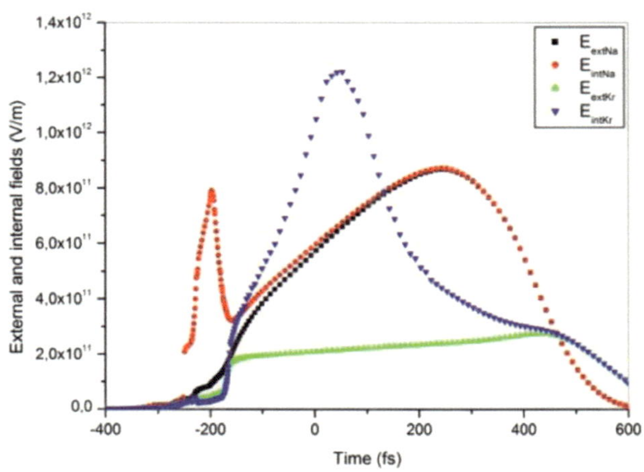

Figure 2.
*Time dependence of the internal and external electric fields for the Na and Kr clusters contained 2×10^7 atoms per cluster. Same parameters of the laser are used as those in **Figure 2**.*

cluster; the expansion of the cluster lowers the electron density to bring the system into resonance with laser field. The electron density in the cluster drops to $3n_{crit}$ (**Figure 3**), and we have the second resonance; (when the ratio $n_e/n_{crit} = 3$), at −200 fs and 47 fs for both clusters Na and Kr respectively. We see that times of resonances are different for both clusters, metallic cluster (Na) and rare gas cluster (Kr).

The field inside the cluster is strongly enhanced and reached the values 1.22×10^{12} V/m and 8.00×10^{11} V/m for Kr and Na clusters, respectively (**Figure 2**); then we have a very rapid deposition of the energy into the electrons; at this point, the ions very rapidly stripped the hot electrons. The internal field $E_{int}(t)$ is almost equal to the external field when the electron density n_e is very low.

The calculated time evolutions of the hydrodynamic pressure and the coulomb one are shown in **Figure 4** for both clusters. The total charge on the cluster increases, and the repulsive forces lead to the increase of the coulomb pressure to

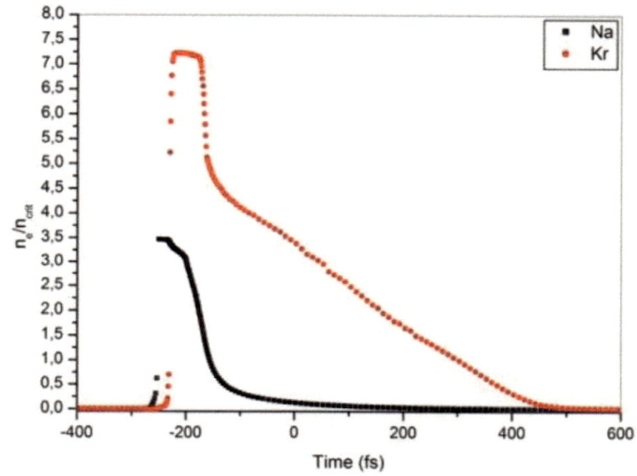

Figure 3.
Calculated electron density n_e normalized by the critical electron density n_{crit} as a function of time for 2×10^7 atoms Na and Kr clusters irradiated by an intense femtosecond laser with a pulse duration of 200 fs, wavelength of 390 nm, and peak intensity of 10^{17} W/cm².

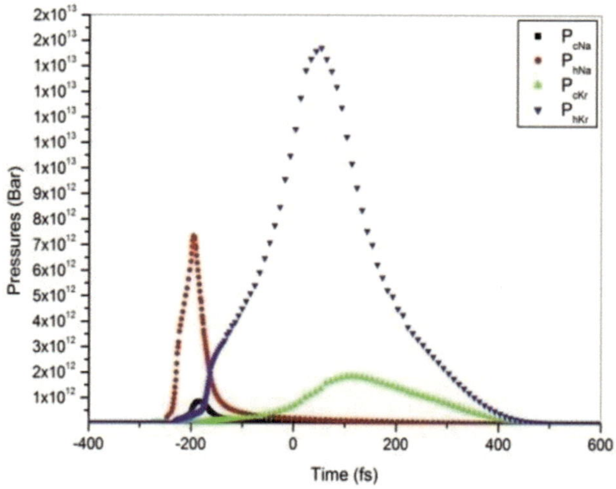

Figure 4.
Calculated time variation of the coulomb and hydrodynamic pressures for Na and Kr clusters (contained 2×10^7 atoms) irradiated by an intense femtosecond laser with a pulse duration of 200 fs (FWHM), wavelength of 390 nm, and a peak intensity of 10^{17} W/cm².

2.0×10^{12} Bar and 9.9×10^{11} Bar for Kr and Na clusters, respectively. The values of the coulomb pressure for both clusters are small compared to the hydrodynamic pressure due to the hot electrons which are 1.5×10^{13} Bar for Kr cluster and 7.5×10^{12} Bar for Na clusters. This pressure leads to the increase of the cluster expansion velocity. During the majority of time, the dominant pressure is the hydrodynamic pressure with a little contribution of the coulomb explosion force; then the hydrodynamic pressure is dominant in driving the explosion. It has been noticed that for small diameter clusters, the coulomb explosion force plays a key role [5]. Our study of the coulomb explosion of large Kr and Na clusters shows that the hydrodynamic pressure plays an important role than the coulomb one like the previous experimental results [28].

When electrons gain energy through inverse Bremsstrahlung collisions, the time evolution of ion energy (**Figure 5**) roughly follows the evolution of the internal field. The increase of the electronic temperature leads to higher ionized states, and electron-free streaming rate increases sharply; then electrons with high energy can leave the cluster (**Figure 6**), as the maximum electron energy is above 3 and 6 keV for Na and Kr clusters, respectively.

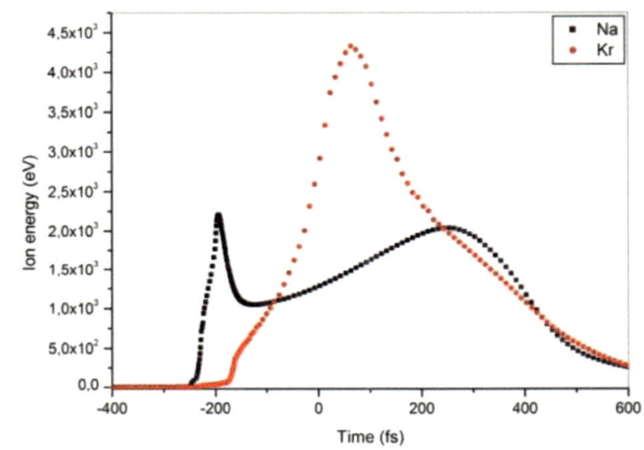

Figure 5.
*Time evolution of ion energy for Na and Kr clusters. The laser parameters are the same as in **Figure 4**.*

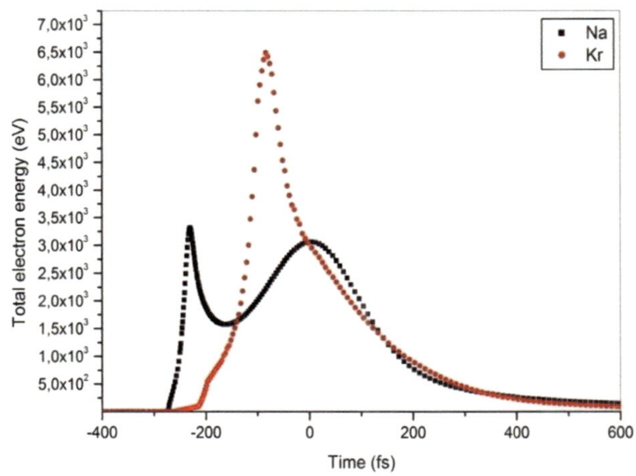

Figure 6.
Total electron energy as a function of time of Na and Kr clusters (contained 2×10^7 atoms) irradiated by a femtosecond laser with a peak intensity of 10^{17} W/cm^2, pulse duration of 200 fs, and wavelength of 390 nm.

The coulomb pressure induced by this loss of the plasma neutrality combined with the hydrodynamic pressure leads to the expansion of the nanoplasma. The enhancement of the internal field (**Figure 2**) leads to a very efficient absorption of the laser energy, resulting to the production of high charge states and high energetic ions 4.4 keV for Kr cluster and 2.2 keV for Na cluster (**Figure 5**). Then the combined effect of free streaming and the coulomb and hydrodynamic pressures leads to the final explosion of the cluster.

4. Conclusions

A modified nanoplasma model has been contributed to examine the cluster explosion dynamics and scaling of ion energies. We found that the excitation of large atomic cluster produces a superheated plasma. We also compared the behavior of the nanoplasma in the Kr rare gas and Na metallic clusters irradiated by an intense femtosecond laser. A similar behavior was found for both cases metallic and rare gas clusters in the time evolution of the different physical parameters (the cluster radius, the electron density, the internal and external electric fields, the pressures, and ion energies). The formation of nanoplasma in the case of Na cluster is rapid than in the case of Kr cluster −255 fs for Na cluster and −230 fs for Kr cluster. We have found that the ion energies in the Kr and Na clusters are 4.4 and 2.2 keV, respectively, which may provide a new ultrahigh energy ion source. The ionization and expansion of the Na metallic cluster are faster than that of the Kr rare gas cluster. The hydrodynamic pressure plays an important role in the interaction with the laser field for both cases (metallic and rare gas clusters); the hydrodynamic pressure was found of order of 10 times than the coulomb pressure; then we conclude that the hydrodynamic pressure is responsible for the dynamic of ionization, expansion, and explosion of the clusters.

Author details

Boucerredj Noureddine* and Khaled Beggas
Semiconductor Laboratory, Physics Department, Faculty of Sciences, Badji Mokhtar Annaba University, Algéria

*Address all correspondence to: boucerredj@yahoo.fr

References

[1] Reinhart PG, Suraud E. Introduction to Cluster Dynamics. Germany: Wiley-VCH Verlag Weinheim; 2004

[2] Nishihara K, Hamitani H, Murakami M, Bulanov SV, Esirkepov TZ. High energy ions generated by laser driven coulomb explosion of cluster. Nuclear Instruments and Methods in Physics Research. 2001;**464**:98-102

[3] Perry MD, Mourou G. Science. 1994; **32**:917

[4] Kundu M, Bauer D. Optimizing the ionization and energy absorption of laser-irradiated clusters. Physics of Plasma. 2008;**15**:033303. DOI: 10.1063/1.2896578

[5] Ditmire T, Tish JWG, Springate E, et al. Nature (London). 1994;**386**:54

[6] Springate E, Hay N, Tish JWG, Mason MB, Ditmire T, Hutchinson MHR, et al. Explosion of atomic clusters irradiated by high intensity laser pulses: Scaling of ion energies with cluster and laser parameters. Physical Review A. 2000; **61**:063201

[7] McPherson A, Thompson BD, Krainov AB, et al. Nature (London). 1994;**370**:631

[8] Lezius M, Dobosz S, Normand D, Schmidt M. Physical Review Letters. 1998;**80**:261

[9] Ditmire T, Donnelly T, Rubenchikh AM, et al. Physical Review A. 1996;**53**:3379

[10] Ditmire T, Zweiack J, Yonovsky VP, et al. Nature. 1999;**398**:489

[11] Ditmire T, Smith RA, Tish JWG, Hutchinson MHR. High intensity laser absorption by gases of atomic clusters. Physical Review Letters. 1997;**78**(6): 3121-3124

[12] Heidenreich A, Last I, Jortner J. Nanoplasma dynamics in Xe clusters driven by ultraintense laser fields. European Physical Journal D: Atomic, Molecular, Optical and Plasma Physics. 2008;**46**:195-202

[13] Arbeiter M, Fennel T. Rare-gas clusters in intense VUV XUV and soft X-ray pulses: Signature of the transition from nanoplasma-driven cluster expansion to coulomb explosion in ion and electron spectra. New Journal of Physics. 2011;**13**

[14] Last I. and Jortner J., Microscopic and hydrodynamic impact energy transfer from nanoplasma electrons to ions in exploding clusters, Molecular Physics. UK: Taylor and Francis; 2018; **116**(19-20):2461-2473. DOI: 10.1080/00268976.2018.14 78135

[15] Micheau S, Jouin H, Pons B. Modified nanoplasma model for laser cluster interaction. Physical Review A. 2008;**77**:053201

[16] Last I, Schek I, Jortner J. The Journal of Chemical Physics. 1997;**107**:6685

[17] Ditmire T. Simulation of exploding clusters ionized by high intensity femtosecond laser pulses. Physical Review A. 1998;**57**(6):R4094-R4097

[18] Megi F, Belkacem M, Bouchen MA, Suraud E. Zweicknagel, on the importance of damping phenomena in clusters irradiated by intense laser fields. Journal of Physics B: Atomic, Molecular and Optical Physics. 2003;**36**:273-282

[19] Ditmire T, Springate E, Tisch JWG, et al. Explosion of atomic clusters heated by high-intensity femtosecond laser pulse. Physical Review A. 1998;**57**(1):369

[20] Boucerredj N, Beggas K, Djabri I. Coulomb explosion of Kr clusters irradiated by intense laser field. In: 2nd International Meeting on Materials for Electronic-Applications (IMMEA) May 8–10. Tunisia: Hammamet; 2009

[21] Boucerredj N, Beggas K, Djabri I. Study with modified nanoplasma model the interaction of intense laser with large metallic clusters of Pb. In: International Conference in Nanomaterial and Renewable Energy (ICNMRE) July 5–8. Morocco: Esafi; 2010

[22] Ammosov MV, Delone NB, Krainov VP. Tunnel ionization of complex atoms and of atomic ions in an alternating electromagnetic field. Soviet Physics - JETP. 1987;**64**:1191

[23] Lotz WZ. Electron-impact ionization cross-sections and ionization rate coefficients for atoms and ions from hydrogen to calcium. Physica A. 1968; **216**:241

[24] Tish JWG, Hay N, Mendham KJ, Springate E, Symes DR, Comley AJ, et al. Interaction of intense laser pulses with atomic clusters: Measurements of ion emission, simulations and applications. Nuclear Instruments and Methods in Physics Research. 2003;**205**: 310-323

[25] Landau LD, Lifshitz EM. Electrodynamics of Continuous Media. Oxford: Pergamon; 1984

[26] Jakhson JD. Classical Electrodynamics. New York: John Wiley and Sons; 1975

[27] Djaoui A, Offenberger AA. Heating of underdense plasmas by intense-short pulse lasers. Physical Review E. 1994;**50**: 4961-4968

[28] Polishchuk AY, Vehn JMT. Electron-ion relaxation in a plasma interacting with an intense laser field. Physical Review E. 1994;**49**:663-666

[29] Johnston TW, Dawson JM. Correct values for high-frequency power absorption by inverse bremsstrahlung in plasmas. Physics of Fluids. 1973;**16**:722

[30] Spitzer L. Physics of Fully Ionized Gases. New York: Interscience; 1967

Chapter 8

Imaging in Low Back Pain

Haider N. Al-Tameemi

Abstract

Medical imaging plays an important role in the evaluation of low back pain (LBP). The choice of certain radiological method over other depends on many factors like patient's presentation, presence of contraindication, availability, relative cost of the test, and the expected impact of the results on management. Radiological evaluation helps the physician reach the most likely cause of LBP, confirm the provisional diagnosis, provide alternative one, or narrow the differential diagnosis. Plain X-ray radiograph is useful in initial general assessment. Magnetic resonance imaging (MRI) is the imaging modality of choice in the evaluation of LBP because of elegant demonstration of anatomical details and many pathologies. Computerized tomography (CT) can provide high-resolution images of the bony structures and is particularly invaluable in trauma. Other imaging modalities are rarely used usually as problem-solving or in selected conditions. For example, sonography may have a role in the evaluation of soft tissue lesions and the sacroiliac joints. Angiography is useful for vascular evaluation. Isotope imaging may be used in the elucidation of of hidden cause of pain (tumors or fracture). Conventional myelography and discography are virtually obsolete in current clinical practice because of the presence of much safer and accurate new modalities. Finally, interventional radiology has an increasing role in treating certain conditions.

Keywords: imaging, radiology, X-ray, MRI, CT scan, sonography, angiography

1. Introduction

The main role of imaging in patients with low back pain (LBP) is to help physicians reach the most likely cause of the pain. The use of one or more of the different radiological investigations in a suitable manner will enhance the detection of the underlying cause of the LBP in a timely way. Moreover, radiology has a role not only in the diagnosis but also in the treatment of some conditions that lead to LBP. The rapidly advancing interventional radiology is increasingly utilized as an adjuvant or sole therapeutic option of a variety of conditions like vascular malformations and tumors.

There are many imaging modalities that can be used in the evaluation of LBP. Selection of the appropriate modality depends on different factors, like the patient's conditions, clinical state, availability, and cost of the test and the presence of certain contraindications.

We will discuss each imaging modality from different points of view starting with the most to the least commonly and widely used ones.

IntechOpen

2. Imaging modalities used in evaluation of LBP

2.1 Plain X-ray radiograph

Plain X-ray radiograph is a simple radiological examination that can be used to give initial general picture. Generally it has a limited role in the evaluation of acute LBP [1]. Fluoroscopy (dynamic X-ray) has an important role in guiding interventional procedures and both diagnosis and pain management [2].

2.1.1 Advantages

1. Low cost (much cheaper than CT and MRI)

2. Availability (it is usually readily available in almost all hospitals and, most of time, in centers, including emergency and far health centers)

3. Noninvasiveness (it involves no risky intervention)

4. Acceptable resolution of bony structures of the spine and pelvis

5. No significant contraindication (apart from pregnancy)

2.1.2 Limitations

1. Poor visualization of soft tissue structures.

2. Cannot show the details of the spinal canal.

3. Cannot show the intervertebral disc material and hence will not give information about the type nor the severity of herniation [3].

4. Factors like obesity and excessive bowel gases may obscure some abnormalities or make interpretation of X-ray difficult.

5. Source of radiation to the patient.

2.1.3 Techniques

After appropriate positioning of the patient (usually supine, sometimes prone or lateral), the examination is performed in not more than a few seconds in general.

2.1.4 Diagnostic value

Generally, plain X-ray can show:

1. General alignment of the spine, any asymmetry, or gross deformity. Straightening of the normal lumbosacral curvature may be considered as an indirect sign of acute spasm or pain.

2. Abnormal position of certain vertebra in the form of foreword (spondylolisthesis) or backward (retrolisthesis) shifting of the vertebral body relative to the vertebrae above and below.

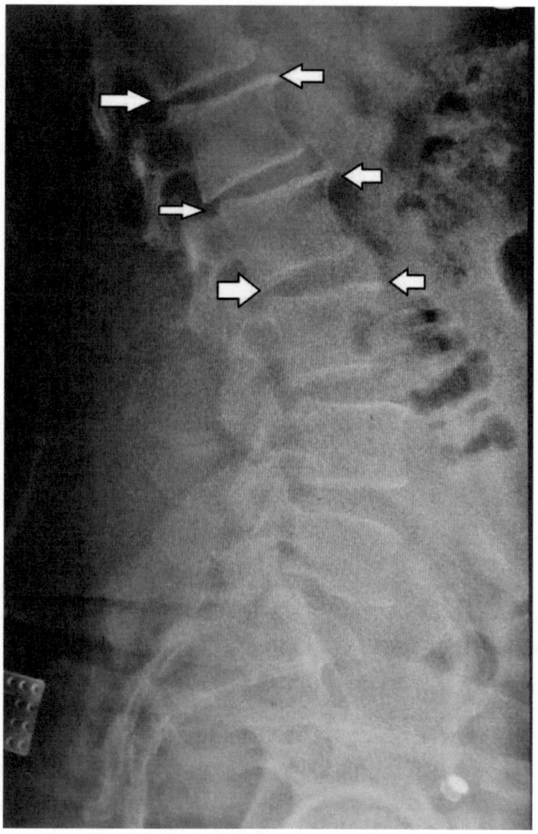

Figure 1.
Lateral X-ray radiograph of the lumbosacral spine of a patient with LBP showing narrowing of multilevel intervertebral disc spaces with marginal osteophytes (white arrows) and juxta-endplate sclerosis, consistent with spinal degeneration.

3. Some congenital anomalies of the spine like scoliosis, spina bifida (defect in the back of spine), and vertebral deformities in the form of abnormal shape, number, and alignment.

4. Lumbosacral spinal degenerative changes in the form of disc space narrowing, marginal osteophytes (bony protrusions at the margins of vertebra), and bony sclerosis near disc (**Figure 1**).

5. Signs of spinal infections (disc space narrowing with irregularity, bone destruction, vertebral compression, soft tissue swelling).

6. Diseases of sacroiliac joints like ankylosing spondylitis and Reiter's disease in the form of joint space narrowing, irregularity, erosions, and sclerosis.

7. Bone tumors (benign and malignant) affecting the spine or pelvis in the form of lytic (black) or sclerotic (white) lesions.

2.2 Magnetic resonance imaging (MRI)

Since its first clinical application in medicine about four decades ago, magnetic resonance imaging (MRI) has revolutionized the imaging evaluation of the human body particularly the musculoskeletal system, brain, and spine. In patients with LBP, MRI is considered as the imaging modality of choice because of its excellent demonstration of both the anatomy of the lower back and most of regional pathologies [4].

2.2.1 Advantages

1. No radiation

2. Noninvasiveness

3. Excellent imaging quality of the spine and pelvis

4. Most accurate modality in the demonstration of the neural structures that can be the source of LBP like the spinal cord and nerve roots

2.2.2 Limitations

1. Presence of contraindications (see Section 3.5).

2. Imaging quality is affected by the patient's motion, so examination needs a cooperative and stable patient. Agitated and unstable patients result in bad images that are diagnostically not useful.

3. Long examination time, usually 10-20 minutes (longer than X-ray and CT scan).

4. Not so accurate in certain conditions like cortical fractures and calcification (CT scan and lesser extent X-ray radiography are better).

5. Clinical MRI mismatch. Many abnormal findings specially disc degeneration or mild herniation are seen incidentally on spinal MRI of asymptomatic persons and vice versa where MRI can be completely normal in the presence of significant LBP. This has raised the issue of importance of clinical and electrophysiological correlation and the controversial need for more specific diagnostic tests like discography [5].

2.2.3 Contraindications

1. Cardiac pacemakers.

2. Metallic object (shells, bullet, orthopedic fixation devices). However, if the fixation devices used were made of MRI-compatible metals (more expensive than regular ones), MRI examination can be safely done despite some artifacts that slightly reduce diagnostic imaging quality.

3. Early pregnancy. However, the chemical material sometimes may be injected intravenously (gadolinium) to enhance the images, and this should not be used throughout pregnancy.

4. Claustrophobia. Seen in about 5–10% of population, when the person is unwillingly afraid of being in closed space. This problem may be resolved by reassurance and careful description of the procedure to the patient. Sometimes the patient will need to be examined using a special type of MRI device called "open type" that has wider aperture and more space around the patient. Rarely sedative/antianxiety drugs may be prescribed.

2.2.4 Techniques

The patient is asked to remove any removable metallic objects like rings and watches. The examination usually takes 10–20 min to complete according to the clinical indication and requested sequences. MRI examination is usually composed of two or more of "sequences." Body tissues and pathologies appear differently on each sequence, and the most commonly used are T1-weighted and T2-weighted sequences. Other sequences used are Short tau inversion recovery (STIR) (suppresses fat signal in the bone marrow and fatty areas), diffusion-weighted image (DWI) (to assess certain features of some lesions according to water molecules diffusion) and fluid attenuation inversion recovery (FLAIR) (suppresses signal of the fluid). Sometimes, when vascular assessment is important, magnetic resonance angiography (MRA) is also conducted. Spinal pathologies can cause both structural and morphological changes. Degeneration of the intervertebral disc, for instance, can be diagnosed when the normal high signal (white) of the disc is lost (disc appears black). Disc prolapse is diagnosed when a part of the disc is seen outside the normal contour of the disc bulging into the spinal canal or compressing the nerve roots within the intervertebral canal.

2.2.5 Diagnostic value

From the diagnostic point of view, MRI can help in:

1. Comprehensive evaluation of the lower back (lumbosacral region) in which the most common causes of LBP arise. Variable pathologies from degenerative, neoplastic, infective, and congenital abnormalities can be elegantly demonstrated. Important and common causes of LBP like intervertebral disc herniation, nerve root compression, lumbar canal stenosis, and degeneration of the small joints at the back of the spine (facet or apophyseal joints) and intraspinal ligaments (ligamentum flavum) can be seen (**Figure 2**).

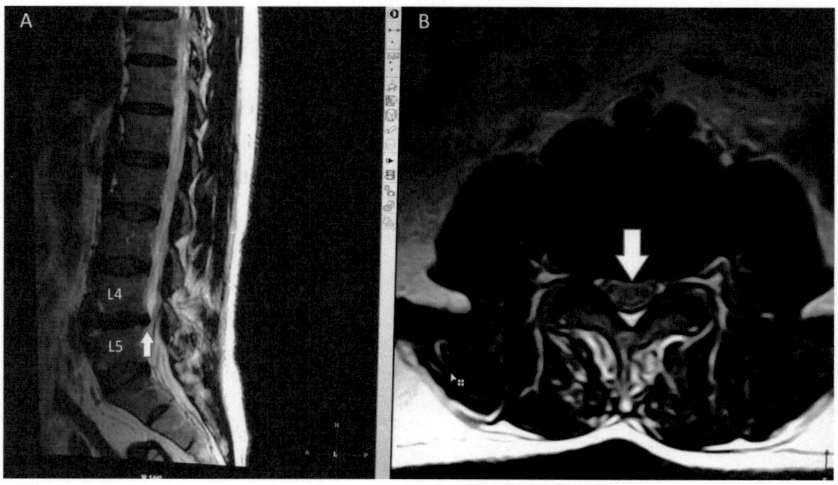

Figure 2.
Sagittal (A) and axial (B) T2-weigted MRI images of the lumbosacral spine of a patient with LBP showing hypointense (dehydrated) intervertebral disc between the L4 and L5 vertebrae with posterior protrusion (white arrows) causing bilateral neural canal narrowing and root compression. Note the normal signal (hyperintense with hypointense cleft) of other discs.

2. Complete evaluation of the spinal canal including the presence of any narrowing, stenosis or abnormal widening. Measurement of the cross-sectional area of the canal is preferred to objectively assess stenosis.

3. Detection of any mass within or outside the spinal canal that may have effects on the spinal cord, nerve roots, soft tissues, or bones including vertebral bony tumors, nerve sheath tumors, lipomas, vascular malformations, etc. [6].

4. Accurate depiction of variable congenital spinal malformations that may present as LBP, like scoliosis, spina bifida, myelomeningocele, dermoids, splitting of the cord, low-lying tethered spinal cord, and absence of the sacrum.

2.3 Computerized tomography (CT) scan

The main physical principle underlying CT scan is X-radiation. The use of CT scan in the assessment of the patients with LBP is limited to certain conditions that either need further evaluation after MRI, patients who cannot be examined by MRI, or when the clinical scenario necessitates CT from the start (like trauma).

2.3.1 Advantages

1. Excellent depiction of the bony anatomy and bony changes at the lower back including lumbosacral spine and pelvis [7]. Bony changes can be readily seen on CT even before appearance on X-ray radiographs.

2. Can compete with MRI in the visualization of lumbar disc prolapse and bony spinal canal stenosis.

3. Relatively more widely available and accessible than MRI.

4. Lesser cost than MRI.

5. Less affected by the patient's motion during examination than MRI.

6. Short time of examination (usually less than minute).

7. Can be safely done for patients with metallic implants, shells, and magnetic fixation devices (but some image compromise).

2.3.2 Limitations

1. High risk of radiation exposure [7]. This is indeed the most important disadvantage of CT scan. CT scan delivers a huge amount of radiation dose to the patient (almost 100 times that of chest X-ray).

2. Poor demonstration of intraspinal neural content like cord and nerve root and early bone marrow infiltration [8].

3. Artifacts from metallic fixation devices or shells may badly affect the quality of image and can obscure some anatomical and pathological findings.

2.3.3 Techniques

The patient is routinely examined in supine position (back on the table), and the exposure done while the patient is asked not to move. The examination is relatively rapid (less than minute).

2.3.4 Diagnostic value

Considering a case of LBP, CT scan can show:

1. Abnormal bone densities of the lower spine and pelvis including osteolytic (blacker) and osteoblastic (whiter) lesions, with a lot of possible causes, from incidental nonsignificant (bone islands), benign (hemangiomas, osteomas) to malignant (metastases, multiple myeloma) lesions.

2. Abnormal configuration and deformity, whether of congenital origin or as sequel of old trauma or surgery.

3. Traumatic findings like linear fracture, partial or complete vertebral compression, and burst vertebra (**Figure 3**). CT scan can detect even tiny fractures of the bony cortex particularly of the posterior spinal elements (laminae and pedicles) which are usually difficult to be seen and may be missed on X-ray radiographs [3].

4. Size, shape, and exact location of the metallic foreign body or shell or bullet.

5. Bone erosion secondary to inflammation, infection, or tumor.

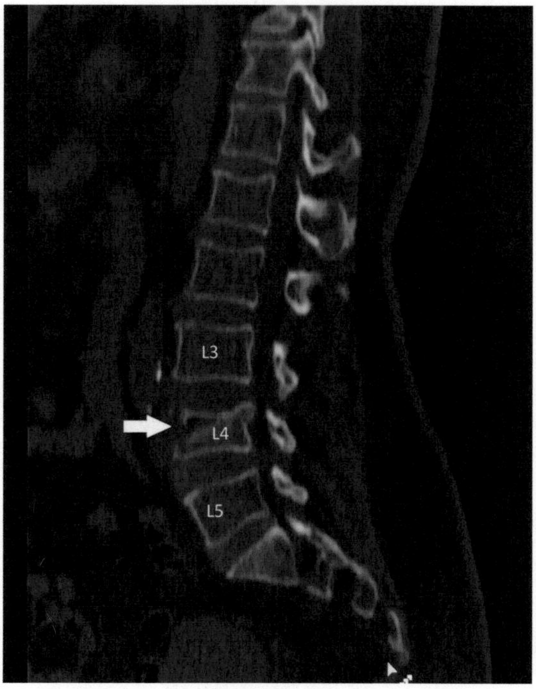

Figure 3.
Sagittal reconstructed image (bone window) of lumbosacral spinal CT scan of a patient with LBP after trauma, showing burst fracture of L4 vertebra.

6. Many (but not all) features of spinal degeneration like intervertebral disc space narrowing, marginal osteophytes, and facet joints sclerosis.

7. Calcification. CT scan is the best imaging modality in defecting calcification in paraspinal soft tissues, ligaments, muscles, or within a mass. The presence of calcification has important diagnostic impact as it helps narrow the differential diagnostic list or may reach final diagnosis like hydatid cyst or para-articular calcification in some chronic arthropathies.

2.4 Sonography

Sonography (or ultrasound scan, echography) utilizes sound waves of very high frequency that are normally not audible by human beings. Its application in the evaluation of patients with LBP is limited because the ultrasound waves are badly affected by tissues like the bone and air.

2.4.1 Advantages

1. Cheap

2. Readily available (almost everywhere, anytime), repeated easily

3. Generally safe (no exposure to ionizing radiation)

4. Noninvasiveness

5. No significant risk or contraindication

2.4.2 Limitations

1. Highly operator-dependent. Examination in general and of the musculo-skeletal system in particular needs expertise and adequate training. Both false-positive and false-negative results are common by inexperienced examiners.

2. Highly device-dependent. The quality of the ultrasound machine has high impact on the quality and subsequent diagnostic outcome of the examination, particularly of the musculoskeletal ones. Many advanced technologies (both software and hardware) are currently emerging that help improve diagnosis which are typically supplied to the mid or even only high-level (expensive) machines.

3. Demonstrates only selected anatomical details of lower back.

2.4.3 Techniques

No special preparation is usually required. Routinely, a gray-scale (B-mode) scan is applied. Color Doppler scan is used when the assessment of blood flow and vascular imaging is required. Advanced techniques like 3D, 4D, and elastography are increasingly utilized both in research and specialized centers in musculoskeletal imaging.

2.4.4 Diagnostic value

Using different technologies, ultrasound can help physician improve the management of the LBP in both diagnostic and therapeutic aspects:

1. Can easily show the content and vascularity of any superficial soft tissue mass that may be related to the LBP.

2. Can show some details of some congenital anomalies at the lower back like cysts or fatty (lipomas, dermoid) and vascular (hemangiomas) lesion.

3. Can show some abnormal features of sacroiliac joints like effusion, bone surface erosion, and soft tissue swellings [9].

4. It may help detect some pelvic conditions that may be directly or indirectly related to the cause of LBP including gynecological condition (inflammatory, tumors, cysts, etc.), bowel masses, and other peritoneal pelvic lesions.

2.5 Isotope imaging

Isotope imaging involves the administration of certain radioactive material into the body (usually intravenously), and then images are created according to the metabolic activity of the targeted tissue/organ.

It usually provides additional diagnostic information to other "structural" imaging modalities (like CT or MRI) to narrow the differential or reach the final diagnosis. Considering the LBP, the most commonly useful isotope imaging modalities used are bone scan and posterior emission tomography (PET).

2.5.1 Advantages

1. Provision of highly useful information about the "metabolic" or "physiological" activity of the lesion, which cannot be obtained by any of the conventional imaging like X-ray, CT scan, or MRI.

2. Can detect hidden causes of pain originating from the bony components of the lower spine and pelvic like stress fracture or metastases in patient with known cancer [10].

2.5.2 Limitations

1. Not cheap examination.

2. It is not readily available as it is commonly present in specialized centers; PET scanners required very sophisticated measures to prepare the isotope material on site just before examination.

3. It is a source for radiation to the patients as well as to medical personals.

4. Has very limited spatial resolution so that interpretation should be correlated with anatomical images. Therefore, most of currently used devices are merged with the conventional anatomical modalities resulting, for example, in hybrid or fusion PET/CT or PET/MRI.

2.5.3 Techniques

Depending on the purpose of the examination, the isotope material (usually incorporated with another carrier substance) is administrated to the patient. The patient is then imaged by the device.

2.5.4 Diagnostic value

1. Generally, metabolically active lesions appear of higher signal than adjacent tissues, so-called "hot spots," and vice versa, inactive lesions have lower signal, so-called cold spots.

2. Can provide invaluable decision about the nature of a vague or suspicious lesion, for example, in the pelvis or lower spine, and whether it is significant and metabolically active or not by assessing its activity.

3. Helps detect many lesions that are metabolically active (like cancer, active inflammation, healing fractures, postoperative) and that are not well or difficult to be seen by the common modalities (CT scan and MRI) [9, 11].

4. Monitoring the response of cancer after radiotherapy and/or chemotherapy and assessing the presence of a residual/recurrent tumor mass after surgical resection.

2.6 Angiography

Many methods are utilized in angiography (visualization of the vascular system), commonly by injecting a contrast material into the vessels and then taking images. In the assessment of patient with LBP, angiography generally has very limited role like in further evaluation or treatment of a vascular lesion [12].

2.6.1 Advantages

1. Provision of details about of the vascular component of the lesion

2. Opportunity of the therapeutic option as some lesions can be treated directly by the angiography by, for example, injecting a special material to occlude and "embolize" the feeding vessel of the mass or vascular anomaly. This is useful for patients who cannot undergo the surgery and with difficult or risky surgical access.

2.6.2 Limitations

1. Not cheap.

2. Not widely available.

3. Needs expertise.

4. High radiation exposure both to patient and staff.

5. Risk of contrast material particularly the iodine-based material (in conventional angiography and CT angiography (CTA)) like allergy and renal damage.

These are much less in MRI (gadolinium) but this carries risk of "systemic progressive sclerosis."

2.6.3 Techniques

Conventional angiography involves injection of iodine-based contrast material through an arterial catheter under fluoroscopy. CTA is less invasive, involving the injection of the contrast material intravenously and the patient is imaged by CT scanner. MR angiography can be done both with and without the use of intravenous chemical contrast material (gadolinium).

2.6.4 Diagnostic value

1. Can help reaching the final diagnosis of a vascular lesion (whether vascular tumor, vascular malformation, hemangiomas)

2. Provides a road map which can be of great value to the surgeon before operating on certain mass by showing detailed vascular anatomy of the region and vascular component of certain mass

3. Has a great value in the therapy of certain vascular lesions and tumors by either treating or reducing the bulk of mass

4. Can also help alleviating the pain in advanced cancer that is not responding to medical treatment and beyond surgical treatment specially in those in end-stage disease

2.7 Myelography

The visualization of the spinal canal lumen is called myelography. It is usually done as a complimentary to other imaging modalities in the evaluation of spinal lesion or canal stenosis.

2.7.1 Advantages

1. More accurate localization of the intraspinal lesion

2. Provision of additional delineation of the spinal canal stenosis

3. Improvement in the diagnosis of some types of nerve root compression

2.7.2 Limitations

1. The conventional and CT myelography techniques are invasive with risk of infection, pain, and radiation exposure.

2. The image quality and resolution of MR myelography (MRM) are not so high and affected by artifacts and depend on the technique used.

2.7.3 Techniques

This can be achieved invasively by injecting a radiopaque iodine-based contrast material into the cerebrospinal fluid (CSF) space of the spinal canal through a needle inserted at the lower back, and the patient is imaged by either conventional X-ray (conventional myelography) or by CT scan (CT myelography) [10]. With the advent of MRI, it became possible to do myelography noninvasively (without need for any intraspinal injection). This so-called MRM is a great advance in spinal imaging. Nowadays, conventional myelography is mostly superseded by CT myelography which is in turn mainly reserved to those who cannot undergo MRI.

2.7.4 Diagnostic value

The interpretation of myelography is usually done after the complete evaluation of other imaging modalities like X-ray, CT, and MRI. The main diagnostic values are:

1. Filling defects in the myelography can be caused by many intraspinal lesions. Therefore, accurate localization of these lesions whether they are extradural, intradural, extramedullary (outside the cord), or intramedullary (within the cord) has great diagnostic value in reaching the most probable cause.

2. MRM can improve the inter-observer and intra-observer agreement in the assessment of lumbar spinal canal stenosis [13], helping more consistent diagnosis when surgical intervention is considered.

2.8 Discography

Discography means the visualization of the content of intervertebral disc after injection of a radiopaque iodine-based contrast material directly into the disc. It is considered as the most accurate method in deciding which disc level is responsible for the LBP and used as reference test in medical researches. However, because it is a very invasive and painful test with many complications and contraindications (like infection, bleeding and severe spinal compromise), its clinical use now is highly restricted for carefully selected patients (e.g., continuous pain with normal noninvasive imaging or sometimes before surgery) and only under experienced interventionalists [14].

DOI: http://dx.doi.org/10.5772/intechopen.90544

Author details

Haider N. Al-Tameemi

Radiology Unit, Department of Surgery, Faculty of Medicine, University of Kufa, Iraq

*Address all correspondence to: haidern.altameemi@uokufa.edu.iq

References

[1] Kerry S, Hilton S, Dundas D, Rink E, Oakeshott P. Radiography for low back pain: A randomized controlled trial and observational study in primary care. The British Journal of General Practice. 2002;**52**:469-474

[2] Palmer WE. Spinal injections for pain. Management. Radiology. 2016;**281**(3):669-688

[3] Jarvik J, Deyo R. Diagnostic evaluation of low back pain with emphasis on imaging. Annals of Internal Medicine. 2002;**137**:586-595

[4] Ciricillo SF, Weinstein PR. Lumbar spinal stenosis. The Western Journal of Medicine. 1993;**158**:171-177

[5] Amundsen T, Weber H, Lilleas F, et al. Lumbar spinal stenosis. Clinical and radiologic features. Spine. 1995;**20**:1178-1186

[6] Frassica FJ, Khanna JA, McCarthy EF. The role of MR imaging in soft tissue tumor evaluation: Perspective of the orthopedic oncologist and musculoskeletal pathologist. Magnetic Resonance Imaging Clinics of North America. 2000;**8**(4):915-927

[7] Tins B. Technical aspects of CT imaging of the spine. Insights Imaging. 2010;**1**:349-359

[8] Tins BJ, Lalam RK, Cassar-Pullicino VN, et al. Bone metastases. 2: Pelvis and appendicular skeleton. In: Davies AM, Sundaram M, James SL, editors. Imaging of Bone Tumors and Tumor-like Lesions. Berlin Heidelberg: Springer; 2009. pp. 346-365

[9] Ghosh A, Mondal S, Sinha D, Nag A, Chakraborty S. Ultrasonography as a useful modality for documenting sacroiliitis in radiographically negative inflammatory back pain: A comparative evaluation with MRI. Rheumatology. 2014;**53**:2030-2034. DOI: 10.1093/rheumatology/keu220

[10] Johnson AW, Weiss CB Jr, Stento K, Wheeler DL. Stress fractures of the sacrum. An atypical cause of low back pain in the female athlete. The American Journal of Sports Medicine. 2001;**29**:498-508

[11] Prandini N, Lazzeri E, Rossi B, et al. Nuclear medicine imaging of bone infections. Nuclear Medicine Communications. 2006;**27**:633-644

[12] Guzman R, Dubach-Schwizer S, Heini P, Lovblad KO, Kalbermatten D, Gerhard Schroth G, et al. Preoperative transarterial embolization of vertebral metastases. European Spine Journal. 2005;**14**:263-268. DOI: 10.1007/s00586-004-0757-6

[13] Al-Tameemi H, Al-Essawi S, Shukri M, Naji F. Using magnetic resonance myelography to improve Interobserver agreement in the evaluation of lumbar spinal canal stenosis and root compression. Asian Spine Journal. 2017;**11**(2):198-203. DOI: 10.4184/asj.2017.11.2.198

[14] Walker J, El Abd O, Isaac Z, Muzin S. Discography in practice: A clinical and historical review. Current Reviews in Musculoskeletal Medicine. 2008;**1**:69-83. DOI: 10.1007/s12178-007-9009-9

Chapter 9

Cone-Beam Computed Tomography in Dentomaxillofacial Radiology

Bence Tamás Szabó, Adrienn Dobai and Csaba Dobo-Nagy

Abstract

The daily application of cone-beam computed tomography (CBCT) has been increasing. Not only the number of referrals has been raising, but also the variety of the anatomical regions requested for imaging is also growing in the dentomaxillofacial area. Even though computed tomography (CT) has been widely used in the head and neck region, by the invention of CBCT, some of the drawbacks of CT were overcome and turned into the advantages of the CBCT appliances, such as lower patient dose. In this chapter, we provide a comprehensive picture of the everyday use of CBCT as a modality in the dentomaxillofacial region and its current limitations and expected improvements.

Keywords: cone-beam computed tomography, dental digital radiography, diagnostic imaging, dentomaxillofacial radiology, incidental findings

1. Introduction

The use of cone-beam computed tomography (CBCT) has been increasing in everyday clinical practice. The advantages of CBCT contribute to its spreading not only in the field of dentistry, but also in maxillofacial surgery, otorhinolaryngology, rheumatology, and traumatology. Conventional-computed tomography (CT) has been widely used in the head and neck and other anatomical regions; nevertheless, by the invention of CBCT device for maxillofacial imaging in the 1990s [1], some of the drawbacks of CT were overcome resulting in the application of CBCT as an alternative modality in these regions. CBCT devices offer a compact size, the ability for producing high-resolution volumetric data, and lower patient dose compared to multislice CT (MSCT) [2–4]. Nonetheless, it should be noted that CBCT devices operate in a wide range of dose values [5]; hence, in the particular clinical situation, the proper justification and optimization are crucial. In this chapter, we provide a comprehensive picture of the everyday use of CBCT as a modality in the dentomaxillofacial region and its current limitations and expected improvements.

2. Basic principles of the CBCT

2.1 Image acquisition

The CBCT device consists of an X-ray source and a flat-panel detector, which are connected by a C-arm in a fixed position, but their vertical position can be adjusted according to the anatomical circumstances of the patient. The X-ray source and the detector rotate in the opposite direction from 180° to 360° depending on the exposure setting of the device during the acquisition around the patient's head. The patient is in a standing, sitting, or supine position. In the latter case, the X-ray source and the detector are in a fixed position but rotate in the vertical plane. The average time of the exposure is varying from 5 to 40 s [6] depending on the device and the exposure settings. However, the energy of X-rays generated is usually diverse; hence, not only high but also low energy X-rays leave the X-ray source, which would be absorbed by the soft tissues and would not contribute to the imaging just to the increase of the patient dose. Therefore, low energy X-rays must be absorbed by an aluminum filter or a copper filter [7]. The divergent X-rays emitted from the X-ray tube are usually collimated by lead alloy with a rectangular opening, whose size can be adjusted depending on the size of the exposed volume, i.e., the field of view (FOV). Collimation results in the formation of a cone or pyramidal beam [8], as opposed to conventional CT devices where a fan beam leaves the collimator (**Figure 1**). However, today's MSCTs are using increasingly divergent beams due to the presence of detector rows [9].

During the acquisition, X-rays leave the tube either continuously or in pulse mode, in the latter allowing that the energy passes through the examined volume only at a particular rotation step leading to the reduction of radiation exposure suffered by the patient [7, 10, 11]. Numerous projection images are recorded on the detector during the rotation of the X-ray source and the detector. Unlike conventional CTs, where the patient table is constantly moving during the acquisition, each slice needs to be reconstructed separately to create the entire image data of the volume. The CBCT detector records the incoming X-rays, which were previously passed through and attenuated in various ways by the patient, and transforms them into an electrical signal, which is transmitted to the reconstruction computer. In case of flat-panel detectors (FPDs) such as the thin film transistor (TFT) containing amorphous silicon (a-Si) or

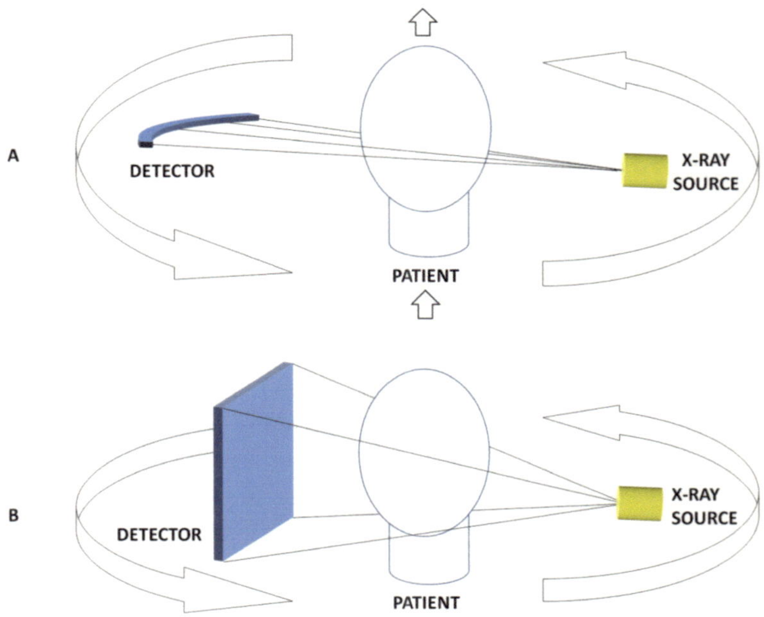

Figure 1.
During the operation of a conventional CT device, the X-ray beam leaving the X-ray tube is fan-shaped and is detected by detector elements arranged in an arc rotating in a direction opposite to the X-ray source around the patient lying on the moving patient table (A). By contrast, CBCT applies a cone-shaped beam during image acquisition, which is recorded by a flat-panel detector (B).

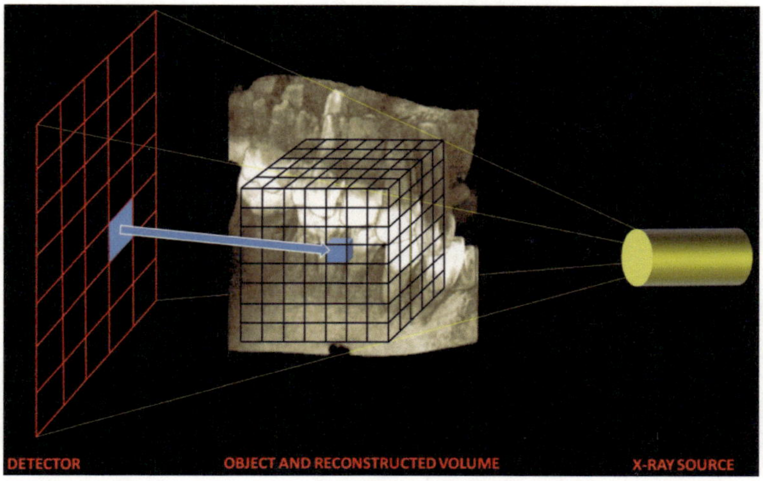

Figure 2.
The reconstruction algorithm projects back a gray value to a voxel according to a particular pixel of the detector on which the intensity of the attenuated X-ray beam was recorded.

complementary metal oxide semiconductor (CMOS), the captured raw image dataset will be appeared as a cylindrical volume after the reconstruction in general [8, 12].

2.2 Image reconstruction

From raw images obtained during the CBCT acquisition, the computer in most cases uses the filtered back projection (FBP) algorithm for CBCT images: the Feldkamp-Davis-Kress (FDK) algorithm [13]. The reconstruction software assigns a grayscale value on a 12-bit scale, in some cases a 16-bit scale, depending on the degree of attenuation (usually the smallest grayscale value corresponds to the air) to each intensity value derived from the linear attenuation of the X-rays passing through the tissues on the pixel matrix of the detector [8]. The FDK algorithm projects back these values to each virtual component of the imaged, namely to each voxel (**Figure 2**). Subsequently, the reconstructed image will be displayed on the monitor as a real volumetric dataset.

3. Artifacts

One of the disadvantages of CBCT is the presence of artifacts that may significantly affect the image quality and interfere with the evaluation of the data [9]. Artifact can be defined as any lesion that is not physically present in the real structure of the imaged volume; nevertheless, it is detectable on the reconstructed image [9, 14]. One of the common artifacts is the beam hardening, which is generated by the polychromatic nature of the X-rays leaving the X-ray tube: the lower energies are absorbed in the tissues with higher absorption capacity; thus, the energy of the X-rays reaching the detector is proportionally higher than its energy emitted just from the X-ray tube. The artifact appears as alternating dark bands and stripes (**Figure 3**) or as a cupping pattern. The latter is explained by the fact that when a homogeneous cylindrical structure is imaged, the energy of the X-rays passing through the center of the volume is "hardened" at a greater level than at the periphery of the object. This leads to a saucer-like shape in the reconstructed image: transparency (i.e., lower grayscale values) in the center area of the scanned object, which decreases steadily as the periphery of the cylindrical structure [15].

Scatter leads to commonly the appearance of dark streaks similar to beam hardening, caused by the interaction between the X-ray and the material of the scanned object. In this case, the original direction of some X-ray photons changes causing computational error in the FDK algorithm. This leads to a decrease in contrast and distortion of grayscale values; however, by increasing the exposure energy and reducing the field of view, scatter can be reduced [7, 9]. It is important to note that selecting a smaller FOV can lead to the "local tomography" effect due to just a part of the object will be counted in the FDK algorithm as the scanned area and the surrounding tissues, which were out of the FOV, will disturb the computational process and the reconstructed image [14].

If a high atomic number material (e.g., metal) is in the path of the X-ray, the element of the detector only records very low intensity value and causes inconsistency during the FDK algorithm. This results in a dark, empty area, or radial streaking in the reconstructed image, which can impair the analysis of the image even on slices further away [9, 16] (**Figure 4**). In English literature, this phenomenon is known as extinction, "missing value," or metal artifacts [8, 14, 15].

The position of the patient during the acquisition has a significant impact on the quality of the reconstructed image data. The FDK algorithm assumes motion-free, constant geometry [9]; therefore, during image acquisition, the changes

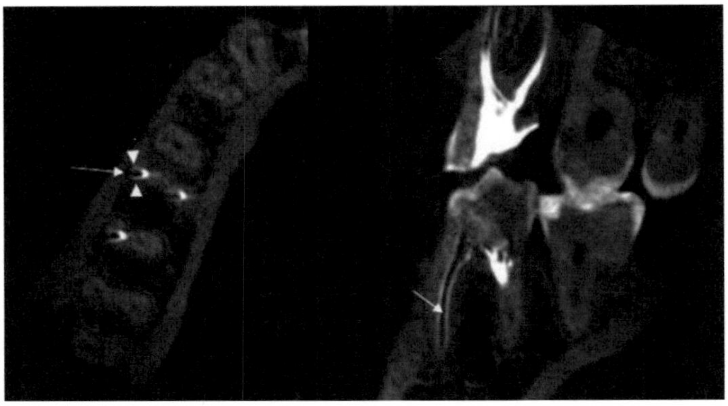

Figure 3.
The white arrows are showing the dark band caused by the artifact along the root filling of the mesiobuccal root canal of the lower right molar tooth, and the white arrows are pointing to the bright bands on the periphery.

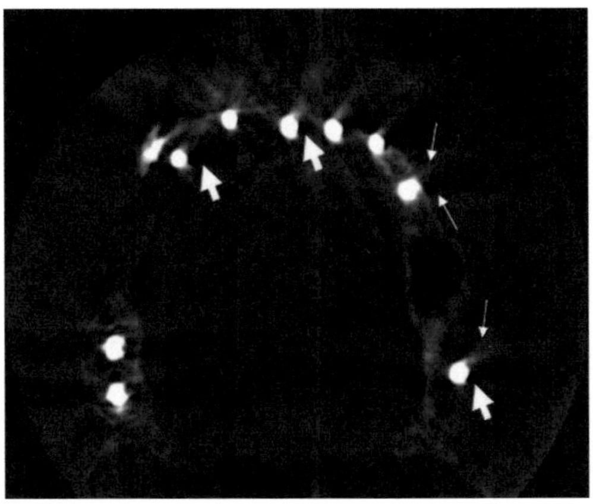

Figure 4.
The thin white arrows are pointing to bright and dark bands caused by beam hardening, while the thick white arrows are pointing to the transparent areas of missing value artifact.

in the patient's position lead to error in the reconstruction algorithm, and the reconstructed images show double contour or blurred image [17], as it is presented on a sagittal slice of our CBCT data (**Figure 5**). Spin-Neto et al. reported that the incidenWce of artifact is higher in CBCT examination of patients under 15 years of age and in small FOV use [18]. Additionally, the patient's displacement of more than 3 mm already causes a significant deterioration of the image quality during exposure [19].

The ring artifact can be described as concentric dark circles on the reconstructed image, mostly on the axial slices (**Figure 6**), indicating the lack of calibration of the detector.

Another possible artifact related to the properties of the CBCT appliance is the aliasing artifact. There are two factors behind: one is the lower frequency of sampling [8], and the other is the geometry of diverging rays emitted from the X-ray tube. This appears on the reconstructed images as lines pointing outward from the

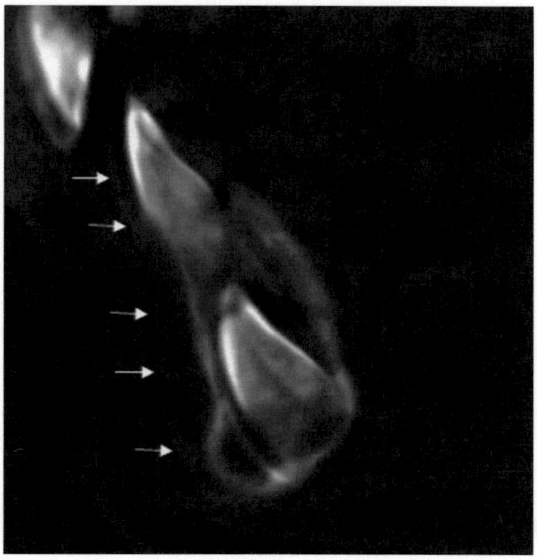

Figure 5.
Motion artifact on a CBCT scan of a 4-year-old patient in the sagittal plane. The white arrows are pointing to the double contour of the blurred image.

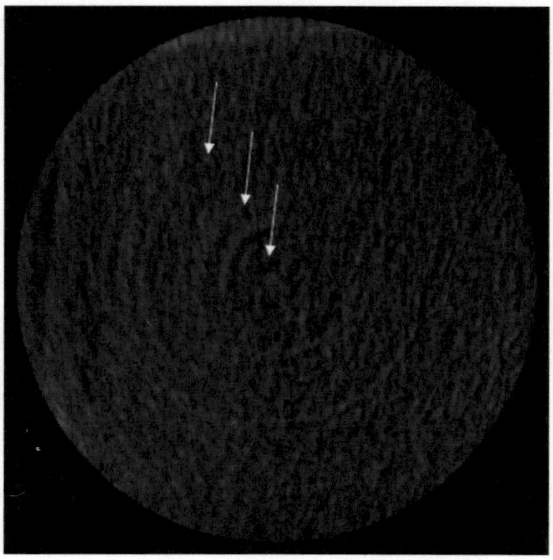

Figure 6.
The white arrows are pointing to the concentric dark circles.

center of the image (Moiré pattern) [14] (**Figure 7**). The artifact can usually be compensated by increasing the number of projections and by the built-in image enhancement algorithms [8, 16].

The essence of the partial volume effect is that if there is a larger difference in the density values (e.g., between the dentin and the root canal) at the edge of the imaged object and this area is close to the boundary of two neighboring voxels, the reconstruction algorithm will only calculate an average grayscale value in relation to the entire voxel (**Figure 8**).

Hence, the algorithm calculates a lower value instead of the real density of the object's boundary [20], so a lower grayscale value appears in this peripheral voxel on the reconstructed image, which might lead to an overestimation of the volume [21].

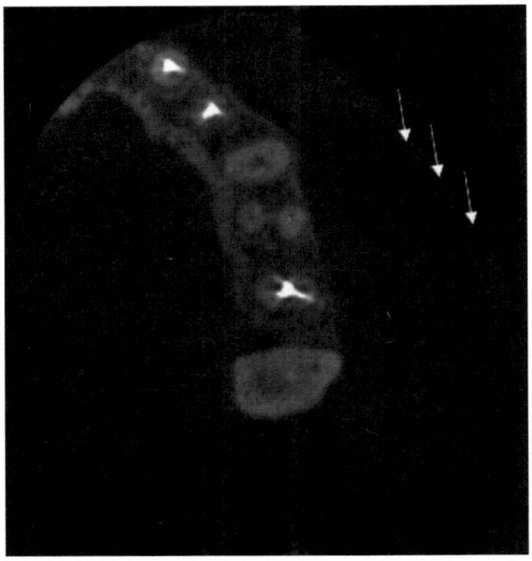

Figure 7.
The white arrows are pointing to the lines appearing farther from the center of the image.

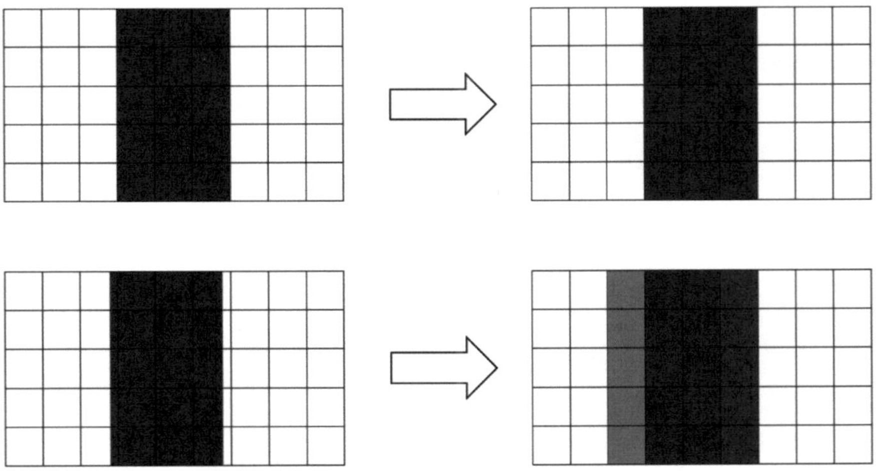

Figure 8.
In the top row, the 3 × 5 pixels in size object "projected" just on the reconstructed pixels, so the real gray scale values of the object are displayed on the reconstructed image. On the bottom row, the object is also 3 × 5 pixels in size but is not "projected" on the reconstructed pixels, so not the actual grayscale values are displayed on the reconstructed image.

4. Imaging of the dentoalveolar region

Since the invention of the CBCT device, numerous publications are concerning the opportunities of CBCT modality in the dentoalveolar region, particularly in endodontics [22–24] and interventions of dentoalveolar surgery [25, 26] as well as orthodontic examinations [27–29]. However, it is important to emphasize that poor image quality can also lead to misdiagnosis and unnecessary exposure to the patient; hence, it is required selecting the appropriate exposure parameters (e.g., FOV, tube voltage, tube current, etc.).

4.1 Dentoalveolar surgery

The basis of the proper diagnosis and successful surgical therapy of a pathological disorder is the comprehensive clinical and radiological knowledge of the anatomical region (**Figure 9**). In cases when panoramic radiographs do not provide sufficient information because of the superimpositions of the anatomical landmarks in the assessed regions, CBCT as a three-dimensional modality shall be considered during the justification. The possible indications of CBCT requirement in this field is well known, such as tooth impaction, pre-implant planning, and surgical guide for implant placement. The proper optimization is essential prior to the exposure including the proper selection of the resolution. The latter shall be applicable at a voxel size of 200 μm for linear measurements, though the accuracy is strongly dependent on other parameters of the used CBCT device [30] and the possible smallest FOV shall be set according to the actual clinical situation [31].

4.2 Endodontics

Dentists should be aware of the root canal system before the endodontic treatment. Currently, commercially available "high-resolution" CBCTs have a nominal voxel size of 100 μm or even 75 μm, which is comparable in size to the apical constriction of a root canal. CBCT as a modality might be a potential choice in selected cases such as complicated root canal systems, when two-dimensional X-ray technique provides limited information [31] (**Figure 10**).

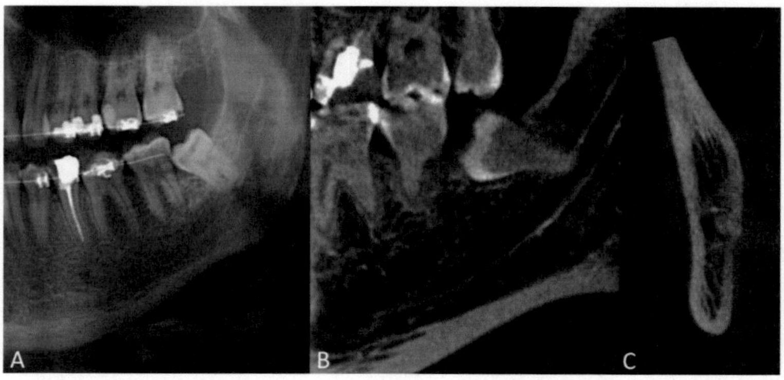

Figure 9.
The relationship of the impacted left lower third molar and left mandibular canal. On the panoramic image, a superimposition is visible (A); however, on the sagittal (B) and coronal (C), CBCT slices the relationship of the apex and the canal is clearly detectable.

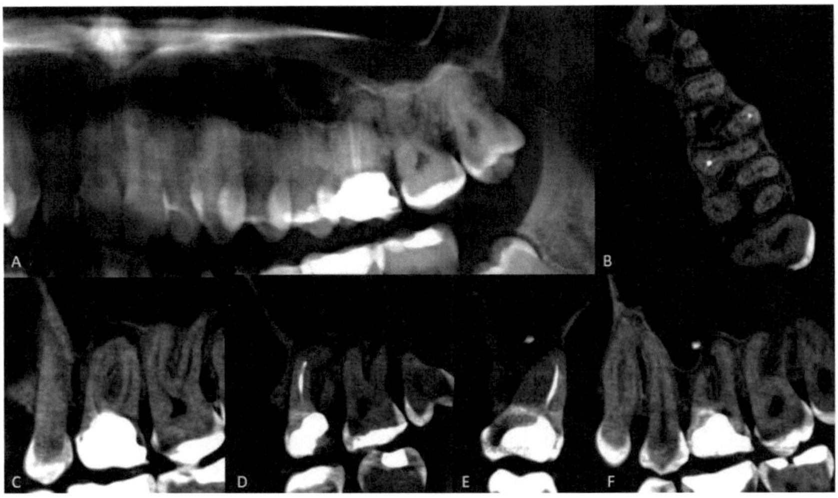

Figure 10.
Insufficient root canal filling of the left upper first molar on the panoramic radiograph (A). The lack of root canal filling of the second mesiobuccal and the distobuccal root canals (B and C). Nearly 90° curvature of the mesiobuccal root and insufficient root canal filling in the first mesiobuccal root canal (D). Hyperdense area in the maxillary sinus potentially due to the extensive use of sealer (E and F). The left upper first premolar is three rooted (B and F).

4.3 Orthodontics

In this field, it is noteworthy to emphasize that the majority of patients are children and young adults [28] who are more sensitive to ionizing radiation. Practitioners must be specifically pay attention to the proper justification and optimization during the orthodontic examinations knowing the fact that CBCT as a modality cannot be used routinely [27].

5. Imaging of anatomical regions apart from the oral cavity

The appearance of CBCT devices has opened new ways of diagnostic radiology not only in the dentistry but also in the field of facial reconstructive surgery, ENT, rheumatology, and orthopedics. Thanks to the lower radiation dose and the accurate imaging of the bony structures, in many cases, CBCT machines mean a proper diagnostic tool for head and neck radiology [2], especially if we consider the high sensitivity of the eye lens.

5.1 CBCT imaging of the paranasal sinuses

All ENT cases make the paranasal sinuses relevant to the anatomical region. One of the most important parameters is the field of view by the CBCT examination of the paranasal sinuses, which should be large enough to include all the sinuses from the frontal sinus to the maxillary sinus (**Figure 11**). Another important rule for the patient's correct position is that the volume should include upper teeth radices to exclude odontogenic sinusitis, which is about 10–12% of all maxillary sinusitis [32].

The frontal and sphenoid sinus, ethmoid cells, maxillary sinuses, and nasal cavity can be studied on the CBCT scan with large field of view. **Figure 1** shows the main anatomical structures of the paranasal sinuses and nasal cavity on CBCT scans.

Although the CBCT imaging has low soft tissue contrast, this method is suitable to detect various signs of inflammation in the paranasal sinuses [33], such as circulated mucosal thickening, air-fluid level, or polypoid lesions, which typical cause widening in the ostiomeatal complex. Furthermore, thanks to visualization of the radices, we can differ the odontogenic and nonodontogenic sinusitis (**Figure 12**). By the malignant lesion, the most important radiological mark is the bone destruction, and in this case, additional imaging techniques, for example, contrast enhanced magnetic resonance imaging (MRI) or CT, are necessary for the accurate diagnosis.

5.2 CBCT imaging of the ear

Nowadays much more CBCT devices are suitable for the examination of the external, middle, and inner ear structures based on high-spatial resolution. For the valuable imaging of the ears, small FOV, low voxel size, approximately 100 μm, and precise patient positioning are required. It is possible to create scan one side or both sides during one rotation, and normally, the whole ear from the skull base to the mastoid process is in the volume tomogram. One of the main advantages of this technique is the low radiation dose oppose to the high-resolution CT, which is an alternative radiologic method to analyze the middle ear ossicles. In the literature, there are some articles that compare reliability of the two techniques for the detection of otosclerosis, but there is no consensus in this question [34–37].

On the CBCT scan, we can identify the external auditory canal, the tympanic cavity with the ossicles [38], and the inner ear structures such as cochlea, semicircular canals, and the facial nerve (**Figure 13**). For the accurate diagnosis, different reconstructions are useful, such as Pöschl reformat for the evaluation of SSC and

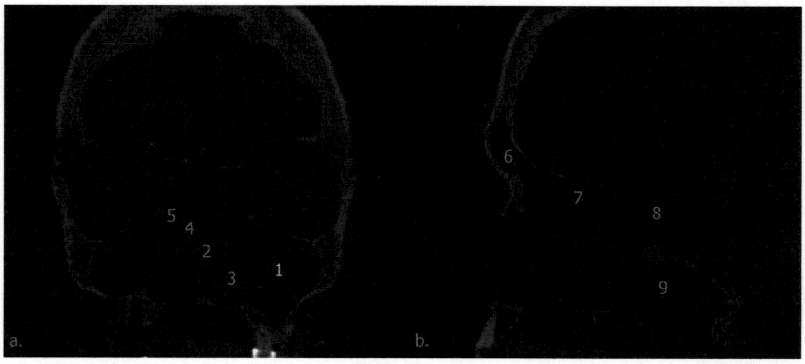

Figure 11.
The main anatomical structures of the paranasal sinuses in coronal (a) and sagittal (b) view (1: maxillary sinus, 2: nasal septum, 3: inferior nasal concha, 4: middle nasal concha, 5: ostiomeatal complex, 6: frontal sinus, 7: ethmoid cells, 8: sphenoid sinus, 9: nasopharynx).

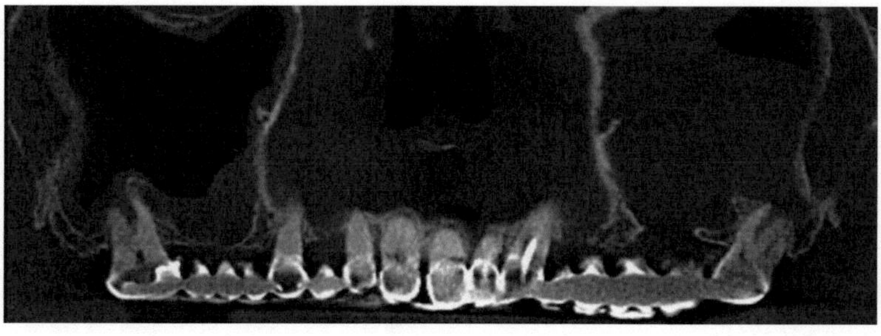

Figure 12.
Sinusitis with oroantral fistula on CBCT scan.

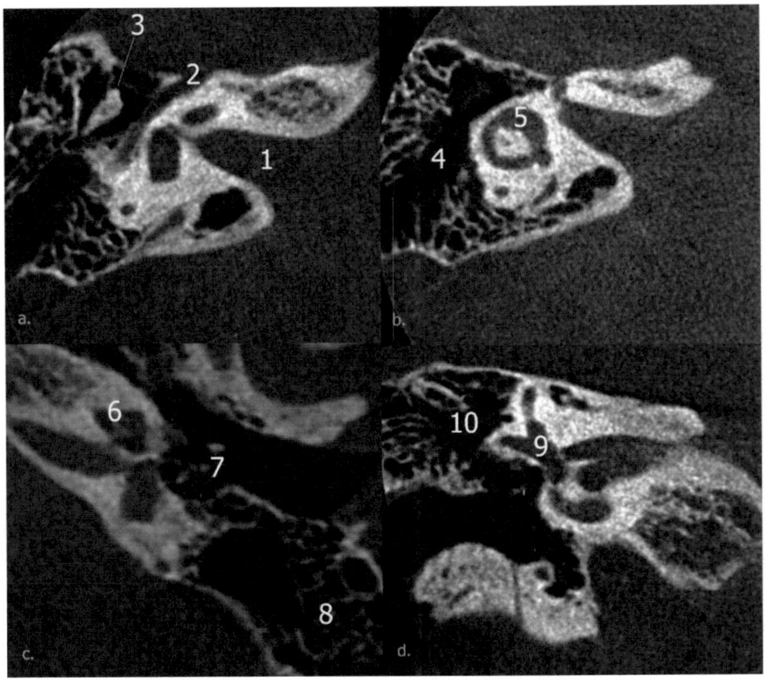

Figure 13.
External, middle, and inner ear structures on CBCT scans (1: internal auditory canal, 2: facial nerve, 3: malleus and incus, 4: tympanic cavity, 5: vestibule and lateral semicircular canal, 6: cochlea, 7: stapes, 8: mastoid cells, 9: semicircular canals, 10: epitympanum).

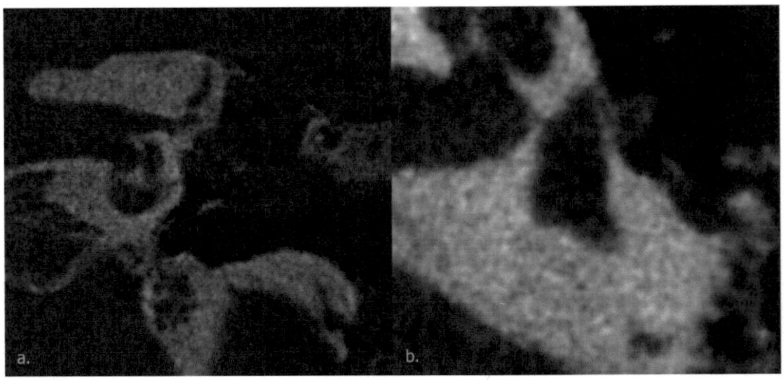

Figure 14.
The image "a" shows cholesteatoma with bone destruction on the tympanic tegmen, and image "b" shows otosclerosis of the stapes.

vestibular aqueduct or Stenvers reformat for the assessment of cochlea, facial nerve, and round window [39].

On the CBCT data, the radiologist can exclude lesions in the external auditory canal, in the tympanic cavity, or in the antromastoidal parts. The most frequently, inflammation can be found in the middle ear cavity or in the mastoid, but special lesions, for instance, glomus tympanicum or cholesteatoma, can be also detected on CBCT images. Despite the poor soft tissue contrast on the CBCT image sequences, there are radiological signs that are characteristic of the above-mentioned lesions. For example, bone destruction and location in the Prussak space are characteristic for the cholesteatoma, and the location on the cochlear promontory is typical for a glomus tympanicum paraganglioma (**Figure 14**).

5.3 CBCT imaging of maxillofacial bones

CBCT imaging is one of the best options for the diagnosis of facial bones, as it can accurately visualize the bone structures and reduce the metal artifact, e.g., from dental implants and crowns, more than conventional CT [40]. The evaluation of the whole facial skull is relevant in the field of the orthognathic and reconstructive surgery; therefore, large field of view is necessary but because of the minimalization of radiation dose larger, e.g., 0.3–0.4 mm voxel size is enough for the correct diagnosis. The CBCT imaging is suitable to detect facial fractures, craniofacial anomalies, bone tumor, or osteomyelitis. Besides of the diagnostics, frequently other aim of this examination is the preoperative surgical planning for orthognathic operation [41], which needs special 3D cephalometric software.

5.4 CBCT imaging of the airway

The CBCT volumetric data with large FOV can be used for the assessment of the upper airways [42, 43]. The volumetric measurement of the airways can facilitate the diagnosis of the patient with obstructive sleep apnea, and it can play an important role in the planning of orthodontic treatment [44, 45].

5.5 CBCT imaging of joints

Most of the CBCT devices are small size devices compared to the conventional CT, by which the patient stands or sits during the exposition. On the market, there are only a few CBCT appliances available in which the patient lies, and the device can scan the joints of the upper and lower extremity and the whole spine. Several constructions of CBCT machines are capable specifically for the joints of upper and lower extremity. These techniques provide the diagnosis of degenerative joint disease, arthritis, tumor-like bone lesion, and fracture of the bones [46, 47].

6. Incidental findings on CBCT image sequences

Frequently in the field of dentistry and ENT large volume, CBCT scans are required for the diagnosis and treatment planning. The increasing of the volume increases the frequency of the incidental findings. Many articles examined the distribution of incidental findings in the dentomaxillofacial CBCT dataset, and the most common incidental findings are the intracranial calcification, frontal hyperostosis, tonsillolith, styloid calcification, antrolith, and artery calcification [48].

6.1 Intracranial calcification

The large volume dental or ENT volume tomogram involves intracranial anatomical structures. One of the most frequent incidental findings is the calcification of the internal carotid artery, in which typical location is by the carotid syphons. The other typical place of the calcification is along the superior sagittal sinus by the cerebral falx.

6.2 Tonsillolith

On the CBCT images, the small calcifications as hyperdense foci are easy to detect; therefore, this imaging technique is capable to visualize small tonsilloliths (**Figure 15**). These calcification clusters can be in or around the tonsils.

These calculi cause no symptoms, and usually, these are only incidental findings, but other pathologies, e.g., foreign bodies, should be distinguished from these stones.

6.3 Styloid calcification

The calcification of the styloid ligament or the elongated styloid process is also known as Eagle syndrome (**Figure 16**), which is usually an incidental finding, but it can cause various symptoms, such as restricted mouth opening, shooting pain in the mandible, pressure in the throat, difficult swallowing, or by the compression on the internal carotid artery or internal jugular vein, and it can lead to intracranial pressure and transient ischemic attack.

6.4 Antrolith

Antroliths are calcifications within the maxillary sinus, which are mostly asymptomatic, and require no treatment (**Figure 17**). On the other hand, larger antroliths may be a symptom of chronic sinusitis or fungal sinusitis, which may involve surgical removal. Antroliths are hyperdense foci or larger masses that are often embedded in the mucoperiosteum in the CBCT images.

6.5 Artery calcification

One of the most frequent incidental findings on CBCT scans is the calcification of the carotid arteries. The calcifications are hyperdensity along the vessels. As the

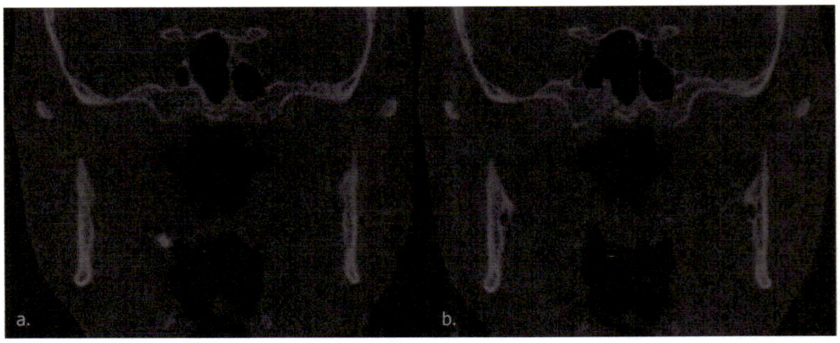

Figure 15.
Tonsillolith as hyperintense foci by the tonsils (a and b).

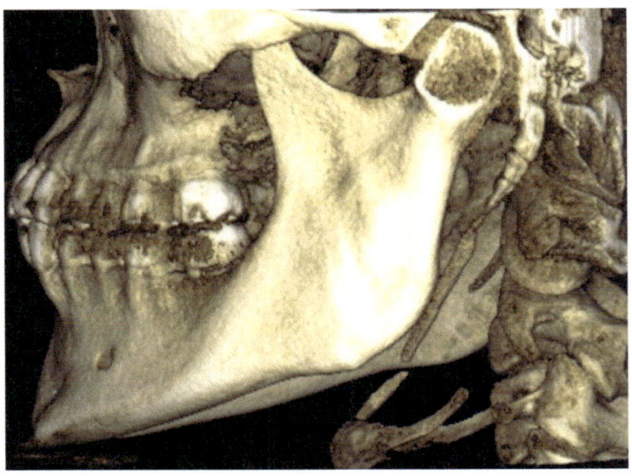

Figure 16.
Calcification of the styloid ligament.

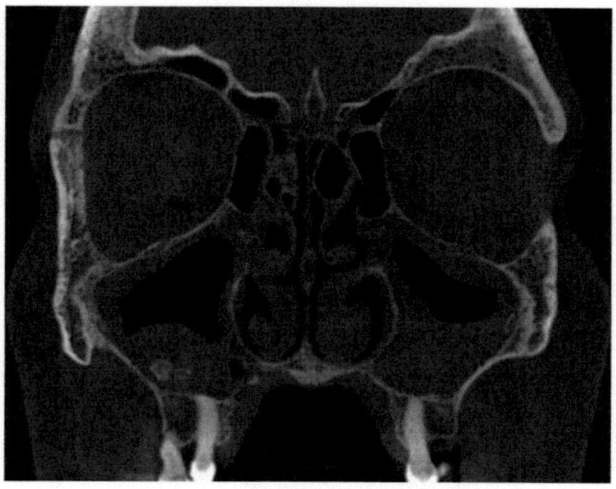

Figure 17.
Calcification in the right maxillary sinus with mucosal thickening.

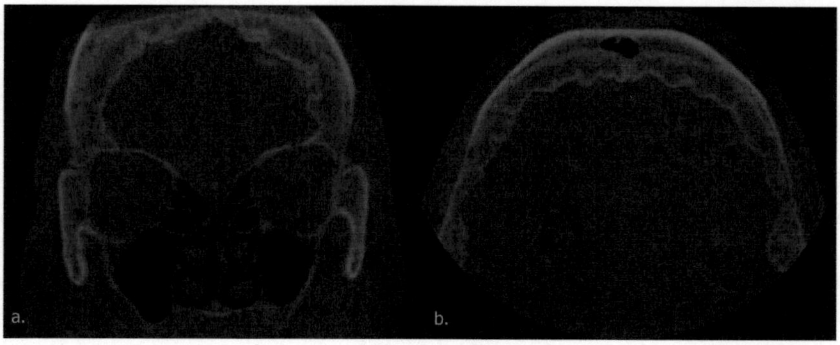

Figure 18.
Hyperostosis frontalis in coronal (a) and axial (b) view.

outcome of the calcification of the internal carotid artery can be stroke, the following stenosis with ultrasound is important.

6.6 Hyperostosis frontalis interna

Hyperostosis frontalis interna is a benign bone formation on the inner table of the frontal bone (**Figure 18**). Normally, it is a symmetric bone thickening, and sometimes, it involves the parietal bones also. The relevance of this disorder is in the differential diagnosis and to distinguish from the pathology such as sclerotic metastases, Paget disease, or fibrous dysplasia.

7. Limitations and expected improvements

One limitation of the present CBCT technology is the high dependence of resolution to the object diameter. This is the reason why this technology is used on head and neck or extremity imaging. The diameter of the chest and abdomen is too big to get higher resolution images than MSCT. Another disadvantage of CBCT is its low quality in soft tissue imaging. Relatively low radiation dose combined with short exposure time provides low quality images of soft tissue structures. This is another reason for CBCT not to use chest and abdomen imaging.

CBCT technology is highly sensitive to object's movement because of the high resolution and of the low number of projection images used for reconstruction.

Usually, it is a challenge to control movements of patient, especially during head imaging. Some positions of the body lower head movements like supine position. Patient movement correction algorithms are available. X-ray scanning combined with video recording the actual movements of patient can be a useful method. Software can analyze and compensate for slight movements gained from video control and provides improved quality images [18].

Filtered back projection based on FDK algorithm is widely used way of reconstruction for CBCT technology. This reconstruction used in wide energy spectrum of X-ray beam arises some artifacts like beam hardening resulting in increased noise of images and misdiagnosis in clinical practice. Beam hardening reduction software is commonly used to improve images; however, this works as an image filtering. In one of our previous study, Monte Carlo analysis was used to compare beam hardening software in reducing beam hardening artifact errors and error by synchrotron micro-CT. Results showed that the beam hardening artifact reduction was comparable with synchrotron micro-CT images [49].

Along with beam hardening artifact scatters (metal artifacts) around the high density metal, implant material frequently causes clinical diagnostic problems since observers unable to detect bone tissue surrounding the implant surface [9]. Soft tissue contrast is further influenced by scatter [50]. Manufacturers usually provide post-processing metal artifact reduction software. Unfortunately this image enhancement is a filter method resulting in loss of peri-implant bone tissue on images [51]. In contrast to maximum likelihood expectation, maximization iterative reconstruction algorithms have proved to improve image quality and to increase signal-to-noise ratio [52]. At present, this is not a widely used method because of the limited capacity of personal computers. However, this way of reconstruction is able to provide image on peri-implant bone tissue. Another way for significant reduction of scatter is dual-energy imaging technology. In a study where single-energy CBCT and dual-energy CBCT were compared, metal artifact with the use of upstream filter on dual energy CT was significantly reduced. This method also resulted in higher signal-to-noise values [53].

Photon-counting CT is a state-of-art technology with the potential to dramatically change clinical 3D imaging. The hybrid-pixel photon-counting detector concept was first demonstrated at CERN in 1991 [54]. The detection of X-ray photons is realized by discriminators, which register a photon only if its energy exceeds a certain threshold value. This way the electronic noise is effectively eliminated and the detector does not affected by dark current signals. Implementation of additional discriminators with different threshold levels to the detector results in discrimination of polychromatic X-ray spectrum into several distinct energy bins [55].

Photon-counting CBCTs are able to reduce radiation dose, reconstruct images at a higher resolution, rectify beam-hardening artifacts, optimize contrast agents' use, and create opportunities for quantitative imaging relative to current CBCT technology [56].

Conflict of interest

The authors declare no conflict of interest.

Author details

Bence Tamás Szabó[1*], Adrienn Dobai[1,2] and Csaba Dobo-Nagy[1]

1 Department of Oral Diagnostics, Faculty of Dentistry, Semmelweis University, Budapest, Hungary

2 Department of Oral Diagnostics, Faculty of Dentistry and Magnetic Resonance Research Center, Faculty of Medicine, Semmelweis University, Budapest, Hungary

*Address all correspondence to: szabo.bence_tamas@dent.semmelweis-univ.hu

References

[1] Farman AG, Scarfe WC. Historical perspectives on CBCT. In: Scarfe WC, Angelopoulos C, editors. Maxillofacial Cone Beam Computed Tomography. Louisville: Springer; 2018. pp. 3-11. DOI: 10.1007/978-3-319-62061-9

[2] Al Abduwani J, Zilin Skiene L, Colley S, Ahmed S. Cone beam CT paranasal sinuses versus standard multidetector and low dose multidetector CT studies. American Journal of Otolaryngology. 2016;**37**:59-64. DOI: 10.1016/j.amjoto.2015.08.002

[3] Brüllmann D, Schulze R. Spatial resolution in CBCT machines for dental/maxillofacial applications-what do we know today? Dentomaxillofacial Radiology. 2015;**44**:20140204. DOI: 10.1259/dmfr.20140204

[4] Nasseh I, Al-Rawi W. Cone beam computed tomography. Dental Clinics of North America. 2018;**62**:361-391. DOI: 10.1016/j.cden.2018.03.002

[5] Ludlow JB, Timothy R, Walker C, Hunter R, Benavides E, Samuelson DB, et al. Effective dose of dental CBCT-a meta analysis of published data and additional data for nine CBCT units. Dentomaxillofacial Radiology. 2015;**44**:20140197. DOI: 10.1259/dmfr.20140197

[6] Pauwels R, Seynaeve L, Henriques JC, de Oliveira-Santos C, Souza PC, Westphalen FH, et al. Optimization of dental CBCT exposures through mAs reduction. Dentomaxillofacial Radiology. 2015;**44**:20150108. DOI: 10.1259/dmfr.20150108

[7] Pauwels R, Araki K, Siewerdsen JH, Thongvigitmanee SS. Technical aspects of dental CBCT: State of the art Dentomaxillofacial Radiology. 2014;**44**:20140224. DOI: 10.1259/dmfr.20140224

[8] Pauwels R. What is CBCT and how does it work? In: Scarfe WC, Angelopoulos C, editors. Maxillofacial Cone Beam Computed Tomography. Louisville: Springer; 2018. pp. 13-42. DOI: 10.1007/978-3-319-62061-9

[9] Kovács M, Fejérdy P, Dobó Nagy CS. A fém műtermék a fejnyaki cone-beam CT (CBCT) képeken. Fogorvosi Szemle. 2008;**5**:171-178

[10] Abramovitch K, Rice DD. Basic principles of cone beam computed tomography. Dental Clinics of North America. 2014;**58**:463-484. DOI: 10.1016/j.cden.2014.03.002

[11] Perényi Á, Bella Z, Baráth Z, Magyar P, Nagy K, Rovó L. A cone-beam komputertomográfia alkalmazása a fül-orr-gégészeti képalkotásban. Orvosi Hetilap. 2016;**157**:52-58. DOI: 10.1556/650.2016.30334

[12] Nemtoi A, Czink C, Haba D, Gahleitner A. Cone beam CT: A current overview of devices. Dentomaxillofacial Radiology. 2013;**4**:20120443. DOI: 10.1259/dmfr.20120443

[13] Feldkamp LA, Davis LC, Kress JW. Practical cone-beam algorithm. Journal of the Optical Society of America. A. 1984;**1**:612-619. DOI: 10.1364/JOSAA.1.000612

[14] Schulze R, Heil U, Grob D, Bruellmann DD, Dranischnikow E, Schwanecke U, et al. Artefacts in CBCT: A review. Dentomaxillofacial Radiology. 2011;**40**:265-273. DOI: 10.1259/dmfr/30642039

[15] Nagarajappa AK, Dwivedi N, Tiwari R. Artifacts: The downturn of CBCT image. Journal of International Society of Preventive & Community Dentistry. 2015;**5**:440-445. DOI: 10.4103/2231-0762.170523

[16] Schulze R, Scarfe WC, Molteni R, Mozzo P, Quality I. In: Scarfe WC, Angelopoulos C, editors. Maxillofacial Cone Beam Computed Tomography. Louisville: Springer; 2018. pp. 95-112. DOI: 10.1007/978-3-319-62061-9

[17] Spin-Neto R, Wenzel A. Patient movement and motion artefacts in cone beam computed tomography of the dentomaxillofacial region: A systematic literature review. Oral Surgery, Oral Medicine, Oral Pathology, Oral Radiology. 2016;**121**:425-433. DOI: 10.1016/j.oooo.2015.11.019

[18] Spin-Neto R, Matzen LH, Schropp L, Gotfredsen E, Wenzel A. Factors affecting patient movement and re-exposure in cone beam computed tomography examination. Oral Surgery, Oral Medicine, Oral Pathology, Oral Radiology. 2015;**119**:572-578. DOI: 10.1016/j.oooo.2015.01.011

[19] Spin-Neto R, Costa C, Salgado DM, Zambrana NR, Gotfredsen E, Wenzel A. Patient movement characteristics and the impact on CBCT image quality and interpretability. Dentomaxillofacial Radiology. 2018;**47**:20170216. DOI: 10.1259/dmfr.20170216

[20] Glover GH, Pelc NJ. Nonlinear partial volume artifacts in x-ray computed tomography. Medical Physics. 1980;**7**:238-248. DOI: 10.1118/1.594678

[21] Ye N, Jian F. Effect of voxel size and partial volume effect on accuracy of tooth volumetric measurements with cone beam CT. Dentomaxillofacial Radiology. 2013;**42**:20130070. DOI: 10.1259/dmfr.20130070

[22] Scarfe WC, Levin MD, Gane D, Farman AG. Use of cone beam computed tomography in endodontics. International Journal of Dentistry. 2009;**2009**:634567. DOI: 10.1155/2009/634567

[23] Michetti J, Maret D, Mallet JP, Diemer F. Validation of cone beam computed tomography as a tool to explore root canal anatomy. Journal of Endodontics. 2010;**36**:1187-1190. DOI: 10.1016/j.joen.2010.03.029

[24] Szabo BT, Pataky L, Mikusi R, Fejerdy P, Dobo-Nagy C. Comparative evaluation of cone-beam CT equipment with micro-CT in the visualization of root canal system. Annali dell'Istituto Superiore di Sanità. 2012;**48**:49-52. DOI: 10.4415/ANN_12_01_08

[25] Van Dessel J, Nicolielo LFP, Huang Y, Coudyzer W, Salmon B, Lambrichts I, et al. Accuracy and reliability of different cone beam computed tomography (CBCT) devices for structural analysis of alveolar bone in comparison with multislice CT and micro-CT. European Journal of Oral Implantology. 2017;**10**:95-105

[26] Parsa A, Ibrahim N, Hassan B, van der Stelt P, Wismeijer D. Bone quality evaluation at dental implant site using multislice CT, micro-CT, and cone beam CT. Clinical Oral Implants Research. 2015;**26**:e1-e7. DOI: 10.1111/clr.12315

[27] De Grauwe A, Ayaz I, Shujaat S, Dimitrov S, Gbadegbegnon L, Vande Vannet B, et al. CBCT in orthodontics: A systematic review on justification of CBCT in a paediatric population prior to orthodontic treatment. European Journal of Orthodontics. 2019;**41**: 381-389. DOI: 10.1093/ejo/cjy066

[28] Kapila SD, Nervina JM. CBCT in orthodontics: Assessment of treatment outcomes and indications for its use. Dentomaxillofacial Radiology. 2014;**44**:20140282. DOI: 10.1259/dmfr.20140282

[29] Dobai A, Vizkelety T, Markella Z, Rosta A, Kucserá Á, Barabás J. Az alsó archarmad Di Paolo-féle vizsgálata Cone-Beam CT adatállományon. Fogorvosi Szemle. 2016;**109**:39-44

[30] Jacobs R, Salmon B, Codari M, Hassan B, Bornstein MM. Cone beam computed tomography in implant dentistry: Recommendations for clinical use. BMC Oral Health. 2018;**18**:88. DOI: 10.1186/s12903-018-0523-5

[31] SEDENTEXCT Guideline Development Panel. Radiation Protection No 172. Cone Beam CT for Dental and Maxillofacial Radiology. Evidence Based Guidelines [Internet]. 2012. Available from: http://www.sedentexct.eu/files/radiation_protection_172.pdf [Accessed: 13 December 2019]

[32] Mehra P, Jeong D. Maxillary sinusitis of odontogenic origin. Current Infectious Disease Reports. 2008;**10**:205-210. DOI: 10.1007/s11908-008-0034-7

[33] Stutzki M, Jahns E, Mandapathil MM, Diogo I, Werner JA, Guldner C. Indications of cone beam CT in head and neck imaging. Acta Oto-Laryngologica. 2015;**135**:1337-1343. DOI: 10.3109/00016489.2015.1076172

[34] Nardi C, Talamonti C, Pallotta S, Saletti P, Calistri L, Cordopatri C, et al. Head and neck effective dose and quantitative assessment of image quality: A study to compare cone beam CT and multislice spiral CT. Dentomaxillofacial Radiology. 2017;**46**:20170030. DOI: 10.1259/dmfr.20170030

[35] Razafindranaly V, Truy E, Pialat JB, Martinon A, Bourhis M, Boublay N, et al. Cone beam CT versus multislice CT: Radiologic diagnostic agreement in the postoperative assessment of cochlear implantation. Otology & Neurotology. 2016;**37**:1246-1254. DOI: 10.1097/mao.0000000000001165

[36] Dahmani-Causse M, Marx M, Deguine O, Fraysse B, Lepage B, Escude B. Morphologic examination of the temporal bone by cone beam computed tomography: Comparison with multislice helical computed tomography. European Annals of Otorhinolaryngology, Head and Neck Diseases. 2011;**128**:230-235. DOI: 10.1016/j.anorl.2011.02.016

[37] Revesz P, Liktor B, Liktor B, Sziklai I, Gerlinger I, Karosi T. Comparative analysis of preoperative diagnostic values of HRCT and CBCT in patients with histologically diagnosed otosclerotic stapes footplates. European Archives of Oto-Rhino-Laryngology. 2016;**273**:63-72. DOI: 10.1007/s00405-015-3490-3

[38] Guldner C, Diogo I, Bernd E, Drager S, Mandapathil M, Teymoortash A, et al. Visualization of anatomy in normal and pathologic middle ears by cone beam CT. European Archives of Oto-Rhino-Laryngology. 2017;**274**:737-742. DOI: 10.1007/s00405-016-4345-2

[39] Reformation in CT of the Temporal Bone [Internet]. 2016. Available from: https://radiologykey.com/reformation-in-ct-of-the-temporal-bone [Accessed: 13 December 2019]

[40] Shintaku WH, Venturin JS, Azevedo B, Noujeim M. Applications of cone-beam computed tomography in fractures of the maxillofacial complex. Dental Traumatology. 2009;**25**:358-366. DOI: 10.1111/j.1600-9657.2009.00795.x

[41] Bai L, Li L, Su K, Bleyer A, Zhang Y, Ji P. 3D reconstruction images of cone beam computed tomography applied to maxillofacial fractures: A case study and mini review. Journal of X-Ray Science and Technology. 2018;**26**:115-123. DOI: 10.3233/xst-17342

[42] Zimmerman JN, Lee J, Pliska BT. Reliability of upper pharyngeal airway assessment using dental CBCT: A systematic review. European Journal of Orthodontics. 2017;**39**(5):489-496. DOI: 10.1093/ejo/cjw079

[43] Souza KR, Oltramari-Navarro PV, Navarro Rde L, Conti AC, Almeida MR. Reliability of a method to conduct upper airway analysis in cone-beam computed tomography. Brazilian Oral Research. 2013;**27**:48-54. DOI: 10.1590/s1806-83242013000100009

[44] Chen H, Li Y, Reiber JH, de Lange J, Tu S, van der Stelt P, et al. Analyses of aerodynamic characteristics of the oropharynx applying CBCT: Obstructive sleep apnea patients versus control subjects. Dentomaxillofacial Radiology. 2018;**47**:20170238. DOI: 10.1259/dmfr.20170238

[45] Obelenis Ryan DP, Bianchi J, Ignacio J, Wolford LM, Goncalves JR. Cone-beam computed tomography airway measurements: Can we trust them? American Journal of Orthodontics and Dentofacial Orthopedics. 2019;**156**:53-60. DOI: 10.1016/j.ajodo.2018.07.024

[46] Jaroma A, Suomalainen JS, Niemitukia L, Soininvaara T, Salo J, Kroger H. Imaging of symptomatic total knee arthroplasty with cone beam computed tomography. Acta Radiologica. 2018;**59**(12):1500-1507. DOI: 10.1177/0284185118762247

[47] Pallaver A, Honigmann P. The role of cone-beam computed tomography (CBCT) scan for detection and follow-up of traumatic wrist pathologies. The Journal of Hand Surgery. 2019;**44**:1081-1087. DOI: 10.1016/j.jhsa.2019.07.014

[48] Barghan S, Tahmasbi Arashlow M, Nair MK. Incidental findings on cone beam computed tomography studies outside of the maxillofacial skeleton. International Journal of Dentistry. 2016;**2016**:9196503. DOI: 10.1155/2016/9196503

[49] Kovács M, Danyi R, Erdélyi M, Fejérdy P, Dobo-Nagy C. Distortional effect of beam-hardening artefacts on microCT: A simulation study based on an in vitro caries model. Oral Surgery, Oral Medicine, Oral Pathology, Oral Radiology, and Endodontics. 2009;**108**:591-599. DOI: 10.1016/j.tripleo.2009.06.009

[50] Tofts PS, Gore JC. Some sources of artefact in computed tomography. Physics in Medicine and Biology. 1980;**25**:117-127

[51] Ferreira JB, Christovam IO, Alencar DS, da Motta AFJ, Mattos CT, Cury-Saramago A. Accuracy and reproducibility of dental measurements on tomogrphic digital models: A systematic review and meta-analysis. Dentomaxillofacial Radiology. 2017;**46**:20160455. DOI: 10.1259/dmfr.20160455

[52] Bechara BB, Moore WS, McMahan CA, Noujeim M. Metal artefact reduction with cone beam CT: An in vitro study. Dentomaxillofacial Radiology. 2012;**41**:248-253. DOI: 10.1259/dmfr/80899839

[53] Iramina H, Hamaguchi T, Nakamura M, Mizowaki T, Kanno I. Metal artifact reduction by filter-based dual-energy cone-beam computed tomography on a bench-top micro-CBCT system: Concept and demonstration. Journal of Radiation Research. 2018;**59**:511-520. DOI: 10.1093/jrr/rry034

[54] Anghinolfi F, Aspell P, Bass K, Beusch W, Bosisio L, Boutonnet C, et al. A 1006 element hybrid silicon pixel detector with strobed binary output. IEEE Transactions on Nuclear Science. 1992;**39**:654-661. DOI: 10.1109/23.159682

[55] Sellerer T, Ehn S, Mechlem K, Duda M, Epple M, Noe PB, et al. Quantitative dual-energy micro-CT with a photon-counting detector for material science and non-destructive testing.

PLoS One. 2019;**14**:e0219659. DOI: 10.1371/journal.pone.0219659

[56] Willemink JM, Pourmorteza A, Pelc JN, Fleischmann D. Photon-counting CT: Technical principles and clinical prospects. Radiology. 2018;**289**:293-312. DOI: 10.1148/radiol.2018172656